METHODS IN MOLECULAR BIOLOGY

Series Editor
John M. Walker
School of Life and Medical Sciences
University of Hertfordshire
Hatfield, Hertfordshire, UK

For further volumes:
http://www.springer.com/series/7651

For over 35 years, biological scientists have come to rely on the research protocols and methodologies in the critically acclaimed *Methods in Molecular Biology* series. The series was the first to introduce the step-by-step protocols approach that has become the standard in all biomedical protocol publishing. Each protocol is provided in readily-reproducible step-by step fashion, opening with an introductory overview, a list of the materials and reagents needed to complete the experiment, and followed by a detailed procedure that is supported with a helpful notes section offering tips and tricks of the trade as well as troubleshooting advice. These hallmark features were introduced by series editor Dr. John Walker and constitute the key ingredient in each and every volume of the *Methods in Molecular Biology* series. Tested and trusted, comprehensive and reliable, all protocols from the series are indexed in PubMed.

Single Cell Analysis

Methods and Protocols

Edited by

Miodrag Gužvić

Department of Urology, University Hospital Regensburg, Regensburg, Germany

Humana Press

Editor
Miodrag Gužvić
Department of Urology
University Hospital Regensburg
Regensburg, Germany

ISSN 1064-3745 ISSN 1940-6029 (electronic)
Methods in Molecular Biology
ISBN 978-1-0716-3620-6 ISBN 978-1-0716-3621-3 (eBook)
https://doi.org/10.1007/978-1-0716-3621-3

This Humana imprint is published by the registered company Springer Science+Business Media, LLC, part of Springer Nature.
The registered company address is: 1 New York Plaza, New York, NY 10004, U.S.A.

Paper in this product is recyclable.

Preface

An exciting time lies ahead for cellular biology. We are finally able to move away from assessing genome, transcriptome, proteome, or metabolome as a mean of a population and instead focus on the features of single cells. Developments in physics, chemistry, and molecular biology have pushed the boundaries of what is possible in cellular analysis to give unprecedented improvements in sensitivity and scale that are allowing us the capability to approach a full functional appraisal of single cells at a rapid rate.

The speed, sensitivity, accuracy, and scope of existing techniques for single-cell detection, isolation, and analysis (e.g., FACS, fluorescence microscopy, dielectrophoresis, whole genome and transcriptome amplification, high-throughput sequencing, etc.) have improved markedly, broadening research horizons, while complementary techniques in organic spectroscopy (e.g., MALDI imaging), inorganic spectroscopy (e.g., ICP-MS), and synchrotron analysis support this with more detailed information on aspects previously overlooked.

This volume summarizes these techniques, their capabilities, and the type of information that can be determined, and aims to give an overview of best practice for implementing them in single-cell analysis in an important and necessary move away from the bulk analysis that is constraining our boundaries.

Multi-omics analysis of single cells is gaining momentum in recent years. Such approaches are somewhat underrepresented in this book, and future editions should more comprehensively cover this important aspect of single-cell analysis. Furthermore, while an attempt was made to make this volume thematically comprehensive, it is somewhat biased toward the analysis of single cancer cells. The future editions of this book should broaden its scope by including protocols on isolation and analysis of single prokaryotic or plant cells. Still, many thematically overlapping or linked protocols presented in this volume enable development of complex and complete workflows to isolate and analyze single cells, even beyond cancer research.

Regensburg, Germany ***Miodrag Gužvić***

Acknowledgments

I am grateful to Rob Hutchinson for support in early phases of preparation of this book, and to contributors and the publisher for their limitless patience, which I tested many times while wrestling many obstacles trying to bring this book to see the light of the day.

This book is dedicated to Christoph A. Klein, who introduced me to the multiverse of single-cell analysis.

Contents

Contributors

BARBARA ALBERTER • *Cellular and Molecular Diagnostics Group, Division of Personalized Cancer Therapy, Fraunhofer Institute of Toxicology and Experimental Medicine ITEM-R, Regensburg, Germany*

BIANCA BEHRENS • *Experimental Surgical Oncology, General, Visceral and Pediatric Surgery, University Hospital and Medical Faculty, Heinrich-Heine-University Düsseldorf, Düsseldorf, Germany*

THOMAS BRAUN • *Department of Cardiac Development and Remodelling, Max Planck Institute for Heart and Lung Research, Bad Nauheim, Germany; German Centre for Cardiovascular Research (DZHK), Partner site Rhein-Main, Frankfurt am Main, Germany*

JOSKA J. BROEKMAAT • *VyCAP B.V, Deventer, The Netherlands*

CATHRYN R. CADWELL • *Departments of Neurological Surgery and Pathology, School of Medicine, University of California San Francisco, San Francisco, CA, USA*

VERA CAPPELLETTI • *Biomarkers Unit, Fondazione IRCCS, Istituto Nazionale dei Tumori di Milano, Milan, Italy*

CLOTILDE COSTA • *Roche-Chus Joint Unit, Translational Medical Oncology Group, Oncomet, Health Research Institute of Santiago de Compostela (IDIS), Santiago de Compostela, Spain; CIBERONC, Centro de Investigación Biomédica en Red Cáncer, Madrid, Spain*

ZBIGNIEW TADEUSZ CZYZ • *Experimental Medicine and Therapy Research, University Regensburg, Regensburg, Germany*

ANDREW D. DUCKWORTH • *Department of Molecular and Clinical Cancer Medicine, Institute of Systems, Molecular and Integrative Biology, University of Liverpool, Liverpool, UK*

GIANCARLO FELICIELLO • *Cellular and Molecular Diagnostics Group, Division of Personalized Cancer Therapy, Fraunhofer Institute of Toxicology and Experimental Medicine ITEM-R, Regensburg, Germany*

CALUM J. GREENHALGH • *Department of Chemistry, Loughborough University, Loughborough, UK*

STEFAN GUENTHER • *Department of Cardiac Development and Remodelling, Max Planck Institute for Heart and Lung Research, Bad Nauheim, Germany*

MIODRAG GUŽVIĆ • *Department of Urology, University Hospital Regensburg, Regensburg, Germany*

SARAH M. HÜCKER • *Fraunhofer Institut für Toxikologie und Experimentelle Medizin, Abteilung Personalisierte Tumortherapie, Regensburg, Germany*

PABLO HURTADO • *Roche-Chus Joint Unit, Translational Medical Oncology Group, Gil Casares Hospital, Oncomet, Health Research Institute of Santiago de Compostela, Santiago de Compostela, Spain*

NAGESH KALAKONDA • *Department of Molecular and Clinical Cancer Medicine, Institute of Systems, Molecular and Integrative Biology, University of Liverpool, Liverpool, UK*

STEFAN KIRSCH • *Fraunhofer Institut für Toxikologie und Experimentelle Medizin, Abteilung Personalisierte Tumortherapie, Regensburg, Germany*

JAKUB KOCHAN • *Faculty of Biochemistry, Biophysics and Biotechnology, Department of Cell Biochemistry, Jagiellonian University, Kraków, Poland*

CÄCILIA KÖSTLER • *Cellular and Molecular Diagnostics Group, Division of Personalized Cancer Therapy, Fraunhofer Institute of Toxicology and Experimental Medicine ITEM-R, Regensburg, Germany*

XIANG LI • *Department of Cardiac Development and Remodelling, Max Planck Institute for Heart and Lung Research, Bad Nauheim, Germany*

AMY J. MANAGH • *Department of Chemistry, Loughborough University, Loughborough, UK*

ALEKSANDRA MARKIEWICZ • *Laboratory of Translational Oncology, Intercollegiate Faculty of Biotechnology, University of Gdansk and Medical University of Gdansk, Gdansk, Poland*

INÉS MARTÍNEZ-PENA • *Roche-Chus Joint Unit, Translational Medical Oncology Group, Gil Casares Hospital, Oncomet, Health Research Institute of Santiago de Compostela, Santiago de Compostela, Spain*

ROSITA MOTTA • *Biomarkers Unit, Fondazione IRCCS, Istituto Nazionale dei Tumori di Milano, Milan, Italy*

LAURA MUINELO-ROMAY • *Liquid Biopsy Analysis Unit, Translational Medical Oncology Group, Health Research Institute of Santiago de Santiago de Compostela (IDIS), Santiago de Compostela, Spain; CIBERONC, Centro de Investigación Biomédica en Red Cáncer, Madrid, Spain*

LISA OOMENS • *VyCAP B.V, Deventer, The Netherlands*

PRAMA PALLAVI • *Department of Surgery, Universitätsmedizin Mannheim, Medical Faculty Mannheim, Heidelberg University, Mannheim, Germany*

THAIS PEREIRA-VEIGA • *Roche-Chus Joint Unit, Translational Medical Oncology Group, Oncomet, Health Research Institute of Santiago de Compostela (IDIS), Santiago de Compostela, Spain; Department of Tumor Biology, Center of Experimental Medicine, University Medical Center Hamburg-Eppendorf, Hamburg, Germany*

ROBERTO PIÑEIRO • *Roche-Chus Joint Unit, Translational Medical Oncology Group, Gil Casares Hospital, Oncomet, Health Research Institute of Santiago de Compostela, Santiago de Compostela, Spain; CIBERONC, Centro de Investigación Biomédica en Red Cáncer, Madrid, Spain*

BERNHARD POLZER • *Cellular and Molecular Diagnostics Group, Division of Personalized Cancer Therapy, Fraunhofer Institute of Toxicology and Experimental Medicine ITEM-R, Regensburg, Germany*

BERNHARD M. POLZER • *Cellular and Molecular Diagnostics Group, Division of Personalized Cancer Therapy, Fraunhofer Institute of Toxicology and Experimental Medicine ITEM-R, Regensburg, Germany*

CAROLINA REDUZZI • *Biomarkers Unit, Fondazione IRCCS, Istituto Nazionale dei Tumori di Milano, Milan, Italy; Division of Hematology/Oncology, Weill Cornell Medicine, New York, United States*

THOMAS SCHAMBERGER • *Experimental Medicine and Therapy Research, University Regensburg, Regensburg, Germany*

STEFAN SCHEUERMANN • *Clinical Health Technologies, Fraunhofer Institute for Manufacturing Engineering and Automation IPA, Mannheim, Germany*

MARCO SILVESTRI • *Biomarkers Unit, Fondazione IRCCS, Istituto Nazionale dei Tumori di Milano, Milan, Italy*

JOSEPH R. SLUPSKY • *Department of Molecular and Clinical Cancer Medicine, Institute of Systems, Molecular and Integrative Biology, University of Liverpool, Liverpool, UK*

MICHIEL STEVENS • *VyCAP B.V, Deventer, The Netherlands*

NIKOLAS STOECKLEIN • *Experimental Surgical Oncology, General, Visceral and Pediatric Surgery, University Hospital and Medical Faculty, Heinrich-Heine-University Düsseldorf, Düsseldorf, Germany*

TENG TENG • *Department of Stem Cell Biology and Regenerative Medicine, University of Southern California, Los Angeles, CA, USA; USC Norris Comprehensive Cancer Center, Keck School of Medicine, University of Southern California, Los Angeles, CA, USA*

PRASHANT V. THAKKAR • *Department of Hematology and Medical Oncology, Weill Cornell Medicine, New York, NY, USA; Department of Medicine, Weill Cornell Medicine, New York, NY, USA*

ARJAN G. J. TIBBE • *VyCAP B.V, Deventer, The Netherlands*

ANDREAS S. TOLIAS • *Department of Neuroscience, Baylor College of Medicine, Houston, TX, USA*

JUSTYNA TOPA • *Laboratory of Translational Oncology, Intercollegiate Faculty of Biotechnology, University of Gdansk and Medical University of Gdansk, Gdansk, Poland*

MARTA VISMARA • *Biomarkers Unit, Fondazione IRCCS, Istituto Nazionale dei Tumori di Milano, Milan, Italy*

MATEUSZ WAWRO • *Faculty of Biochemistry, Biophysics and Biotechnology, Department of Cell Biochemistry, Jagiellonian University, Kraków, Poland*

MICHAIL YEKELCHYK • *Department of Cardiac Development and Remodelling, Max Planck Institute for Heart and Lung Research, Bad Nauheim, Germany*

MIN YU • *Department of Stem Cell Biology and Regenerative Medicine, University of Southern California, Los Angeles, CA, USA; USC Norris Comprehensive Cancer Center, Keck School of Medicine, University of Southern California, Los Angeles, CA, USA; Department of Pharmacology, University of Maryland School of Medicine, Baltimore, MD, USA*

ANNA J. ŻACZEK • *Laboratory of Translational Oncology, Intercollegiate Faculty of Biotechnology, University of Gdansk and Medical University of Gdansk, Gdansk, Poland*

Chapter 1

Single Cell Isolation from Surgically Resected Tissue Via Mechanical Dissociation Using TissueGrinder

Prama Pallavi and Stefan Scheuermann

Abstract

Primary cells form the basis of modern-day in vitro research analysis tools. Many conventional procedures for generating single-cell suspensions from solid tissue are neither robust nor reproducible. Here we describe primary cells isolation from surgically resected tumor tissue via enzyme-free mechanical dissociation using TissueGrinder, a novel semi-automated benchtop device. The isolated cells can be used for any downstream biochemical or cell-based analytic assay.

Key words Primary cells, Enzyme-free cell isolation, Single-cell isolation, Mechanical dissociation

1 Introduction

Primary cells isolated from tissue are very important in vitro tools to answer several complex biological questions. In particular primary cells from surgically resected tissue can provide insight into cellular biological activity and hence new insights into disease [1]. Methods to isolate primary cells from tissues can be broadly divided into two main categories—explant and tissue dissociation techniques using mechanical force and enzymatic digestion [2–5]. For more than 50 years, the explant method dominated the field of tissue culture for obtaining primary cells [6, 7]. In fact, this is one of the simplest techniques: the tissue is finely minced (1–2 mm^3) and placed in a culture flask [8], then the explant is covered with an appropriate cell outgrowth medium. Within a week's time, cells migrate out of the explants and grow on the culture surface [9]. However, success is limited by various factors such as long processing times, low yield, and extensive manual workload, leading to a substantial increase in the overall time from tissue resection to the generation of a cell line. Furthermore, not all types of cells are able to migrate out of explant tissue. The second method, dissociation with enzymes, is not always desired because enzymatic digestion could affect the

Miodrag Gužvić (ed.), *Single Cell Analysis: Methods and Protocols*, Methods in Molecular Biology, vol. 2752,
https://doi.org/10.1007/978-1-0716-3621-3_1,

proteins that might be later needed for labeling or staining, or are required for downstream molecular biological analysis [2]. In the mechanical approach, no additional enzymes or other agents are used to isolate cells from tissue, which could potentially alter the cells. During mechanical dissociation, however, the cells are exposed to a certain level of mechanical stress.

TissueGrinder technology enables the isolation of single cells from biological tissue of any origin by mechanical dissociation, minimizing the mechanical stress to preserve native character of cells to the maximum extent and aims for the utmost biological comparability with the sample donor. With this technology, a combination of shearing and cutting forces is applied to dissociate cells from pre-cut tissue [10]. The duration and speed of shearing and cutting forces can be optimized to develop tissue- and application-specific dissociation protocols. Manually diced tissue sample is transferred into grinding tubes, which contain an integrated grinding gear [10]. After dissociation, centrifugation of the grinding tubes allows easy collection of cells which can be either cultured directly or used for further analytical tests. By combining the process steps of dissociation with filtration and centrifugation in a one-step process, the risk of contamination can be minimized. Thus, TissueGrinder combines tissue-specific dissociation protocols with gentle mechanical dissociation to provide high-quality single-cell suspensions without the use of enzymes.

Due to the high demand for primary cells and the limited availability of tissue samples, the isolation of primary cells should be as efficient as possible. In this method protocol, we describe a workflow for the generation of primary cell cultures using surgically resected tissue obtained from an anal cancer patient via TissueGrinder. The method can be applied also to other cancer tissue types [14].

2 Materials

Clinical tissue specimens are often stored under non-ideal conditions and for long time periods and can therefore be of poor quality at the time of processing. For this reason, it is important to ensure cooling of the specimen after resection and to process it promptly (*see* **Notes**).

2.1 Tissue Transport

1. Prepare centrifuge tube with 30 mL of DMEM medium supplemented with 100 units/mL of penicillin and 100 μg/mL of streptomycin.
2. Prepare a transport container with ice, to transport the tube with medium. Tissue is collected as quickly as possible with a cool box from surgery.

2.2 Dicing of the Tissue

1. 10 cm diameter tissue culture grade petri dish.
2. Scalpels.
3. Cell culture hood.
4. Sterile PBS without Ca^{+2} and Mg^{+2}.
5. Sterile ammonium chloride solution: 0.8% NH_4Cl, 0.1 mM EDTA in water, buffered with $KHCO_3$ to achieve a final pH of 7.2–7.6.
6. Sterile pair of tweezers.

2.3 Tissue Sample Processing with the TissueGrinder

1. Weighing scale with precision of 1 mg.
2. 10 cm diameter tissue culture grade petri dish.
3. TissueGrinder disposable sterile grinding tube set.
4. 1 mL of DMEM medium supplemented with 100 units/mL of penicillin and 100 μg/mL of streptomycin.
5. TissueGrinder device (Fast Forward Discoveries GmbH, Germany).
6. Benchtop centrifuge with swingout rotor available for 50 mL centrifuge tube.
7. Dresden medium: mixture of two thirds of the common DMEM (Sigma-Aldrich) supplemented with 10% FBS (Gibco) and 1% of penicillin (10,000 units/mL)/streptomycin (10,000 μg/mL) solution (Gibco) and one third of growth medium for Keratinocytes (Gibco) supplemented with Bovine Pituitary Extract (25 mg) and EGF Human Recombinant (2.5 μg) (Gibco). This medium was established through previous successes in obtaining cell lines from primary tissue [8].
8. Sterile serological pipettes (10 mL and 5 mL).

2.4 Cell Counting and Viability

1. Luna counting slides.
2. Luna instrument.
3. Sterile Trypan Blue solution 0.4%.

   Cell counting can be also done using Neubauer chamber and light microscope (*see* **Notes**).

2.5 Cell Seeding for Cell Culture

1. Six-well plate tissue culture grade.
2. Culture medium.

   Primary tumor cells isolated from primary tumors are cultured in "Dresden Medium."
3. Sterile serological pipettes (10 mL and 5 mL).

3 Methods

3.1 Tissue Transport

1. Aliquot 20 mL of the DMEM supplemented with 1% of penicillin (10,000 units/mL)/streptomycin (10,000 μg/mL) solution in 50 mL centrifuge tube and place them in 4 °C fridge near operation theater. The amount of medium should be enough to submerge the tissue.
2. Place the surgically resected tumor tissue immediately in the centrifuge tube containing precooled transport medium (4 °C) and transport the tissue on ice from operation room to laboratory and process immediately (*see* **Notes 1** and **2**).

3.2 Pre-Cutting of the Tissue Sample

Note: All these steps should be performed under sterile tissue culture bench (*see* **Notes 3** and **4**).

1. Wash the tumor tissue with PBS supplemented with 1% Penicillin/Streptomycin solution (Penicillin (10,000 units/mL) and Streptomycin (10,000 μg/mL) solution)).
2. If the sample is still blood-soaked, then wash with an ammonium chloride solution to lyse erythrocytes.
3. Pre-cut the tissue into small cubes with an edge length of about 1–2 mm with help of a single-use scalpel.
4. Weigh up to 10–400 mg of the pre-cut tissue pieces in a clean and sterile petri dish.

3.3 Processing with the TissueGrinder

1. Take the disposable sterile grinding tube sets for the TissueGrinder. Assemble by latching the stator into the cell sieve. Next, wash stator and cell sieve with 1 mL of Dresden Medium and transfer the cell sieve to the sterile 50 mL centrifuge tube (*see* **Note 5**).
2. Place 10–400 mg of the prepared tissue pieces in the spare space between the grinding teeth of the rotor (using a 1 mL pipette) using a pair of sterile tweezers. Add 500 μl of the Dresden medium and place the stator with cell sieve on the rotor and close the tube. The detailed assembly of the grinding unit is shown in Fig. 1.
3. Place the closed tube on a TissueGrinder benchtop port and select a predefined tissue dissociation program. There are three different generic programs—soft, medium, and hard—available for soft (easy to cut), medium (mildly fibroblastic), and hard tissues (highly fibroblastic and somewhat calcified). In addition, pre-installed programs for specific tissue types (such as liver, spleen, heart muscle, etc.) can be selected or customized for specific user applications. The generic programs should be selected and executed based on the texture of the

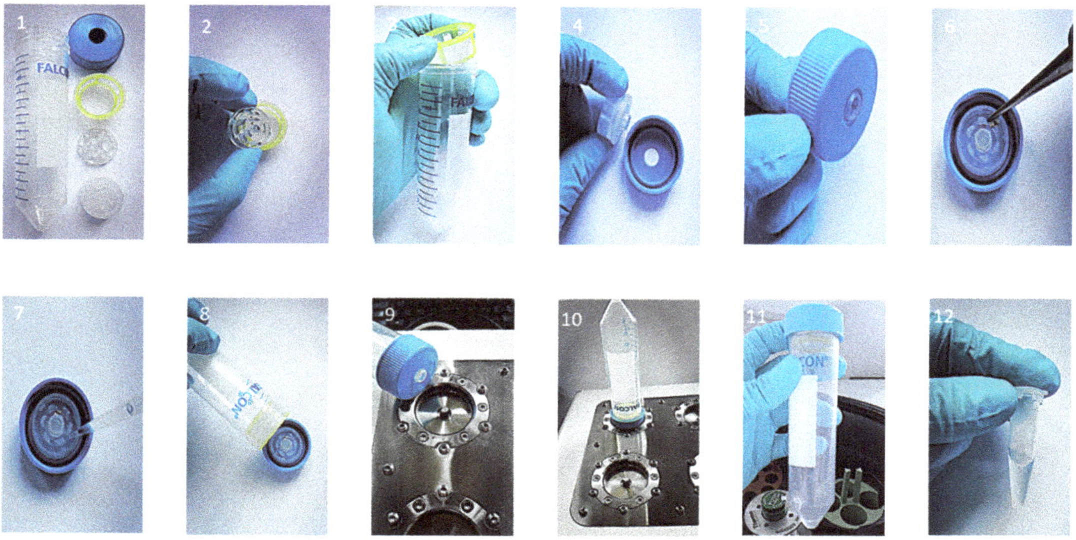

Fig. 1 Assembly of single-use TissueGrinder sterile grinding set (1) and the tissue dissociation workflow (2–12). The grinding set consists of a fitting lid to the 50 mL centrifuge tube with a centered hole, the rotor and the stator both forming the grinding gear, a standard cell strainer with a pore size of 100 μm, and a 50 mL centrifuge tube the assembled grinding set is locked with the rotor's specific notch to its counterpart on the TissueGrinder. The diced tumor tissues, each with edge lengths of 1–2 mm and a total weight of 10–400 mg, are placed in the inner part of the rotor and the unit is assembled as shown under a cell culture bench. Diced tissue fragments are processed into single cells by alternating processes of grinding and cutting which is achieved in the TissueGrinder by clockwise or anticlockwise turning of the rotor. After the dissociation protocol, the tube is transferred directly to a standard laboratory centrifuge and the generated single-cell suspension is centrifuged through the 100 μm cell filter. The cell pellet is resuspended and transferred to a new vessel

tissue. A fine-tuning of the program to suit specific requirements can also be done. Table 1 describes predefined tissue dissociation programs.

4. Since anal cancer tissue is soft, run program soft. Following the program run, centrifuge the tube for 5 min at 350 g (*see* **Notes**/troubleshooting). After centrifugation, open the tube under laminar flow cabinet. Flush the tissue remains with 1 mL Dresden Medium through the cell sieve to ensure that all cells are collected in the 50 mL centrifugal tube. Discard the tissue fragments, which did not pass through the sieve.
5. Remove the cell sieve with the integrated stator from the tube. Close the lid of the tube and centrifuge again for 5 min at 350 g (*see* **Note 6**).
6. Resuspend cell pellet with 3 mL of Dresden Medium. Remove 50 μl of the resuspension in 1.5 mL microcentrifuge tube. This generated single-cell suspension can be used to supply various downstream applications (*see* **Notes 7** and **8**).

Table 1
Steps of three TissueGrinder programs used for processing of tumor tissues

	Program	Soft		Medium		Hard	
Steps	Process	Speed (rpm)	Duration (s)	Speed (rpm)	Duration (s)	Speed (rpm)	Duration (s)
1	Cutting	+30	20	70	30	70	30
2	Grinding	−30	20	−70	30	−70	30
3	Cutting	+30	20	50	25	80	30
4	Cutting	+70	25	−50	25	−80	30
5	Grinding	−70	25			50	15
6	Cutting	+50	25			−50	15

A positive speed means clockwise turns which lead to a cutting of the tissue while a negative speed represents anticlockwise turns which causes a grinding of the tissue. This table is reproduced from [14]. The speed and duration for cutting and grinding can be set for each of the tissue processing programs

Table 2
Representative cell counts after processing tissues from anal cancer sample

Weight of tissue processed per run (mg)	Living cells isolated per 1 mg of tissue
88	1.96×10^5
97	4.460×10^5

3.4 Cell Counting and Viability (see Notes 9–12)

1. Take 10 μl of the cell suspension and mix with 10 μl of the Trypan blue solution.
2. Load 10 μl of this cell solution on to Luna counting slide and insert it in the Luna instrument to count the cells.
3. Note down the cell number (Table 2) and viability of the isolated cells.

3.5 Cell Culture

1. Change the medium of the cells in the six well-plate twice a week until a confluent cell monolayer is observed.
2. Once a monolayer is observed (Fig. 2), sub-culture the cells into T25- and T75-flasks with passage ratio of 1:3 (*see* **Notes 13** and **14**).

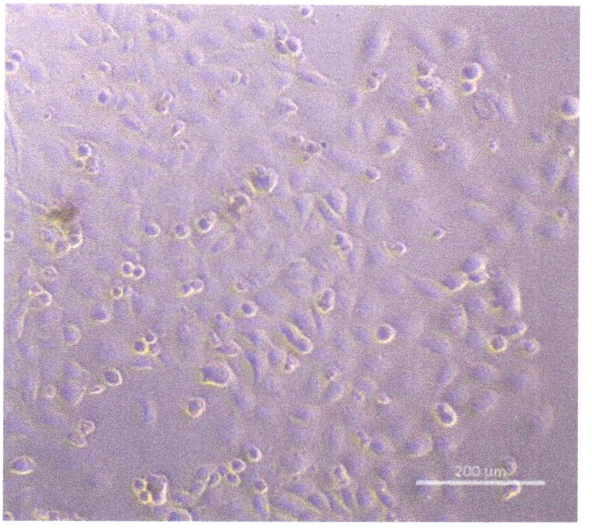

Fig. 2 A representative image showing anal cancer cells isolated via TissueGrinder on the tenth day after processing. The images were taken with 4× Zeiss objective (LDA-Plan421231-991). The white scale bar represents 200 μm

4 Notes

1. A low temperature at around 4 °C helps to slow down metabolic activities that can lead to biochemical changes that alter gene expression, transcription, and the resulting protein profiles. This is especially important for the prevention of cell lysis.
2. Freeze/thaw cycles must also be avoided at this stage as cells can be very sensitive to temperature fluctuations, which can be very dramatic if the starting sample material is very limited.
3. Wear gloves when processing patient-derived tumor samples and avoid closing of scalpel to avoid injuries.
4. Surgically resected tissue from patients suffering from an active HIV-1, or other virologic infectious disease should be processed based on the biological safety level of the laboratory.
5. Sterile grinding kits for TissueGrinder are available with different sieve sizes. A sieve size of 70 μm should be used for cell isolation, while a size of 100 μm can be used to isolate clumps of cells which can be grown in Matrigel for microtissue culture.
6. Cells should always be washed in a buffer solution to remove cell debris, contaminants, and/or medium containing BSA/FCS. In addition, the centrifugation speed should be adapted to the cell type. The generally accepted recommendation for mammalian cells is to centrifuge at 300–900 g for 5 min to avoid cell disruption. Nonetheless, it's crucial to be

aware that excessively low centrifugation speeds can result in the loss of cell pellets and, consequently, sample loss. Thus, great care should be exercised during the washing steps.

Failed single cell analysis: Check cell viability and evaluate cell suspension for total cell yield, cell debris, and aggregates.

7. A multitude of applications and potential exploitation in the field of advanced purification of single-cell suspensions (Fluorescence- or magnetic activated cell sorting (FACS, MACS), laser capture microdissection (LCM), and microfluidic workflows) and single-cell analysis at genomic, transcriptomic, and proteomic levels are emerging [11].
8. Evaluation of a solid tissue-derived single-cell suspension is an important step prior to using cells for, e.g., flow cytometry. The three critical parameters to evaluate after mechanical dissociation of solid tissue are: (i) cell viability, (ii) absence of cell debris, and (iii) absence of aggregates [12]. The viability of cells in a single-cell suspension can be assessed using the trypan blue exclusion assay, in which dead cells incorporate trypan blue into their cytoplasm, while live cells retain their selectively permeable membrane and prevent trypan blue from entering the cytoplasm [13].
9. Cell viability in a single-cell suspension can easily be screened for using the trypan blue cell viability exclusion assay (Neubauer chamber).
10. Cell debris and cell aggregates can also quickly be evaluated using light microscopy.
11. In case of low cell yield, use polypropylene tubes and plates to avoid cell adhesion. Add washing steps.
12. In case of low cell viability, reduce the applied rotational speed of the TissueGrinder protocol by 50% and pipette gently while resuspending.
13. The isolated cells should be monitored every day for the first week to identify contaminated cultures.
14. Medium of the processed cultures should be changed every 3 days.

Acknowledgement

This research work was supported by Wilhelm Müller-Stiftung.

Ethical Statement This study was approved by the local ethics committee of the University Medical Center Mannheim (2012-293 N-MA), and all of the donors gave their written informed consent.

References

1. Lee J-K, Liu Z, Sa JK, Shin S, Wang J, Bordyuh M, Cho HJ, Elliott O, Chu T, Choi SW, Rosenbloom DIS, Lee I-H, Shin YJ, Kang HJ, Kim D, Kim SY, Sim M-H, Kim J, Lee T, Seo YJ, Shin H, Lee M, Kim SH, Kwon Y-J, Oh J-W, Song M, Kim M, Kong D-S, Choi JW, Seol HJ, Lee J-I, Kim ST, Park JO, Kim K-M, Song S-Y, Lee J-W, Kim H-C, Lee JE, Choi MG, Seo SW, Shim YM, Zo JI, Jeong BC, Yoon Y, Ryu GH, Kim NKD, Bae JS, Park W-Y, Lee J, Verhaak RGW, Iavarone A, Lee J, Rabadan R, Nam D-H (2018) Pharmacogenomic landscape of patient-derived tumor cells informs precision oncology therapy. Nat Genet 50(10):1399–1411. https://doi.org/10.1038/s41588-018-0209-6
2. Mitra A, Mishra L, Li S (2013) Technologies for deriving primary tumor cells for use in personalized cancer therapy. Trends Biotechnol 31(6):347–354. https://doi.org/10.1016/j.tibtech.2013.03.006
3. Bols NC, Lee LEJ (1991) Technology and uses of cell cultures from the tissues and organs of bony fish. Cytotechnology 6(3):163–187. https://doi.org/10.1007/BF00624756
4. Gawad C, Koh W, Quake SR (2016) Single-cell genome sequencing: current state of the science. Nat Rev Genet 17(3):175–188. https://doi.org/10.1038/nrg.2015.16
5. Esparza-López J, Martínez-Aguilar JF, Ibarra-Sánchez MJ (2019) Deriving primary cancer cell cultures for personalized therapy. Rev Investig Clin 71(6):369–380. https://doi.org/10.24875/ric.19002832
6. Fischer A (1925) Tissue culture; studies in experimental morphology and general physiology of tissue cells in vitro
7. Parker RC, Morgan JF (1950) Methods of tissue culture. Hoeber Medical Division-Harper & Row, New York
8. Ruckert F, Aust D, Bohme I, Werner K, Brandt A, Diamandis EP, Krautz C, Hering S, Saeger HD, Grutzmann R, Pilarsky C (2012) Five primary human pancreatic adenocarcinoma cell lines established by the outgrowth method. J Surg Res 172(1):29–39. https://doi.org/10.1016/j.jss.2011.04.021
9. Rückert F, Aust D, Böhme I, Werner K, Brandt A, Diamandis EP, Krautz C, Hering S, Saeger HD, Grützmann R, Pilarsky C (2012) Five primary human pancreatic adenocarcinoma cell lines established by the outgrowth method. J Surg Res 172(1):29–39. https://doi.org/10.1016/j.jss.2011.04.021
10. Scheuermann S, Schäfer A, Langejürgen J, Reis C (2019) A step towards enzyme-free tissue dissociation. Curr Dir Biomed Eng 5(1): 545–548. https://doi.org/10.1515/cdbme-2019-0137
11. Hu P, Zhang W, Xin H, Deng G (2016) Single cell isolation and analysis. Front Cell Dev Biol 4:116. https://doi.org/10.3389/fcell.2016.00116
12. Reichard A, Asosingh K (2019) Best practices for preparing a single cell suspension from solid tissues for flow cytometry. Cytometry A 95(2): 219–226. https://doi.org/10.1002/cyto.a.23690
13. Tennant JR (1964) Evaluation of the trypan blue technique for determination of cell viability. Transplantation 2:685–694. https://doi.org/10.1097/00007890-196411000-00001
14. Scheuermann S, Lehmann JM, Mohan RR, Reißfelder C, Rückert F, Langejürgen J, Pallavi P (2021) TissueGrinder, a novel technology for rapid generation of patient-derived single cell suspensions from solid tumors by mechanical tissue dissociation. Front Med 9:721639. https://doi.org/10.3389/fmed2022.721639

Chapter 2

Circulating Tumor Cell Enrichment and Single-Cell Isolation Combining the CellSearch® and DEPArray™ Systems

Cäcilia Köstler, Bernhard Polzer, and Barbara Alberter

Abstract

The analysis of circulating tumor cells (CTCs) has shown potential for detection of cancer spread, prognosis, therapeutic target selection, and monitoring of treatment response. CTCs can be obtained repeatedly by simple blood draws as so-called "liquid biopsy." Thus, they can serve as a surrogate material for primary or metastatic tissue biopsies. In addition, isolation of CTCs provides the possibility to investigate those cells which may hold the (molecular) traits responsible for metastatic progression and ultimately patient death. As such, CTCs represent a target of utmost importance in cancer research and therapy. In this chapter, we describe a workflow for the enrichment of CTCs with the FDA-cleared CellSearch® system followed by the isolation of single CTCs using the DEPArray™ technology enabling further molecular single-cell analyses.

Key words CTC, CellSearch, DEPArray, Single-cell isolation, Dielectrophoresis, Liquid biopsy

1 Introduction

In most cancer patients, CTCs are low in abundance and heterogeneous in morphological and phenotypic profiles [1]. This complicates their enrichment and isolation for subsequent characterization [2]. However, a semi-automated and standardized workflow composed of CTC enrichment with the CellSearch® system and isolation of single CTCs with the DEPArray™ platform produces reliable results and can be applied as a routine workflow [3].

The CellSearch® system for the detection and enumeration of CTCs has been approved for diagnostic use in breast [4], prostate [5], and colorectal cancer [6]. The enrichment of CTCs with this platform makes use of the epithelial origin of carcinoma cells surrounded by mesenchymal cells of the blood compartment. Briefly, Epithelial Cell Adhesion Molecule (EpCAM)-expressing cells are enriched by antibody-coated ferromagnetic beads from a 7.5 mL blood sample and stained for intracellular cytokeratin proteins

Miodrag Gužvić (ed.), *Single Cell Analysis: Methods and Protocols*, Methods in Molecular Biology, vol. 2752,
https://doi.org/10.1007/978-1-0716-3621-3_2,

8, 18, and 19 and for DAPI to determine the presence of nuclear DNA (positive identifiers), as well as for CD45 as a marker for contaminating blood leukocytes (negative identifier). Subsequently, the number of putative CTCs is determined automatically and verified by an experienced operator. Evaluation of CTC numbers at both early- and late-stage disease allows assessment of patient prognosis and is predictive of progression-free survival and overall survival [7, 8].

For research purposes, the precious patient sample analyzed by CellSearch® can be further utilized to isolate CTCs which can then be subjected to molecular analyses at the single-cell level. The DEPArray™ technology has proven beneficial as downstream single-cell isolation procedure. However, a mandatory buffer change before the CellSearch® sample can be transferred to the DEPArray™ cartridge is a critical step in terms of cell loss [2].

The DEPArray™ technology exploits the principles of dielectrophoresis (DEP) and the inducibility of electric dipoles in cells. After polarization of cells, they are trapped one by one in so-called electric “cages” on a chip. At this point, the cells can be detected by microscopic scanning of the whole sample with different channels simultaneously (brightfield and four fluorescence channels). Every event is recorded as an overlay image of the detected object (i.e., putative cell) and can be selected for subsequent separation. This is done by changing the electric field pattern of adjacent cages on the chip making selected cells move forward to recover them either as pools of cells or as single cells.

Once CTCs are isolated as single cells, their nucleic acids can be amplified and a multitude of molecular analyses can be performed, ranging from panel sequencing, copy number variation analysis to genome/exome sequencing.

2 Materials

2.1 CellSearch® System (see Fig. 1), Kits, and Additional Materials Required for Cell Enrichment

2.1.1 CELLTRACKS® AUTOPREP® System

1. CELLTRACKS® AUTOPREP® System User Manual.
2. CELLTRACKS® AUTOPREP® System for enrichment, staining, and collection of the cells of interest.
3. MAGNEST® cartridge mounts.
4. Bottles with caps for waste, instrument buffer, de-ionized water, and cleaning solution.
5. Kit for the monthly maintenance.
6. Uninterrupted power supply.

2.1.2 CELLTRACKS ANALYZER II®

1. CELLTRACKS ANALYZER II® User Manual.
2. CELLTRACKS ANALYZER II® for analysis of the enriched cells.

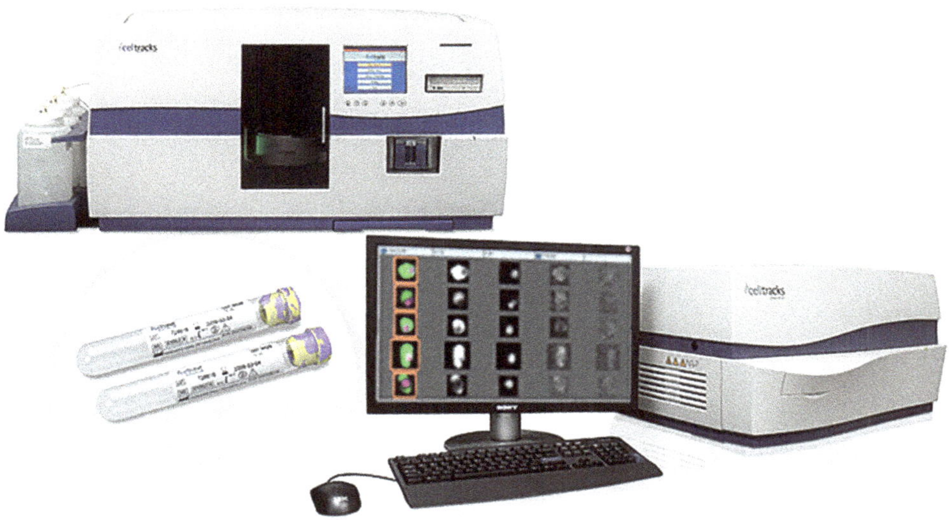

Fig. 1 CellSearch® System consisting of CellSearch® Autoprep System and CellTracks® AnalyzerII. (© by Menarini Silicon Biosystems SpA)

3. Two verification cartridges for the CELLTRACKS® System.
4. CELLTRACKS® justification cartridge.
5. Starter kit for lamp justification.
6. Mercury vapor lamp.
7. Uninterrupted power supply.
8. Local printer for reports.
9. Memorex® DVD + R to back-up results.
10. Fiber-free tissues.

2.1.3 CellSearch® Circulating Tumor Cell Kit

The CellSearch® Circulating Tumor Cell Kit (*see* Fig. 2) is used for capturing CTCs of epithelial origin ($CD45^-$, $EpCAM^+$, and cytokeratins 8^+, 18^+, and/or 19^+). It contains buffers and reagents for 16 enrichments. Up to its use it has to be stored at 2–8 °C (*see* **Notes 1** and **2**). Once it has been opened, the kit can be used for up to 30 days for diagnostic purposes (*see* **Note 3**).

The kit contains the following reagents:

1. 3.0 mL Anti-EpCAM Ferrofluid (brown cap):

   Suspension of 0.022% magnetic particles conjugated to a mouse monoclonal antibody specific for the cell surface marker EpCAM present on epithelial cells in a buffer with 0.03% bovine serum albumin (BSA) and 0.05% ProClin® 300 preservative.

2. 3.0 mL Staining Reagent (white cap):

   0.0006% mouse monoclonal antibodies specific to cytokeratins conjugated to phycoerythrin (PE) and 0.0012% mouse

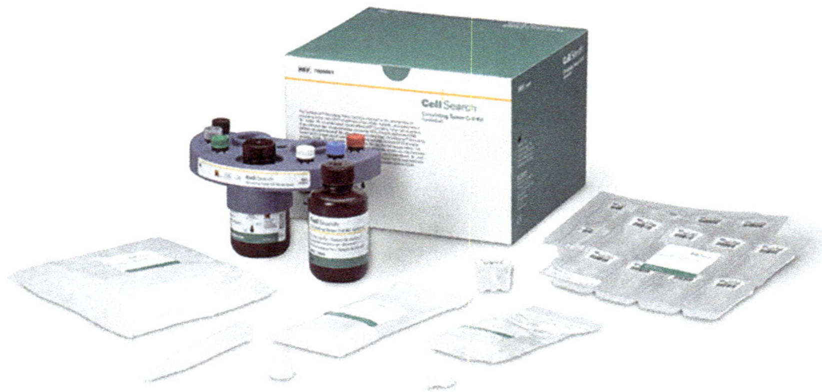

Fig. 2 Content of CellSearch® Circulating Tumor Cell (CTC) Kit. (© by Menarini Silicon Biosystems SpA)

anti-CD45 monoclonal antibody conjugated to allophycocyanin (APC) in buffer containing 0.5% BSA and 0.1% sodium azide.

3. 3.0 mL Nucleic Acid Dye (blue cap):
   0.005% 4′, 6-diamidino-2-phenylindole, dihydrochloride (DAPI), and 0.05% ProClin®300.
4. 3.0 mL Capture Enhancement Reagent (clear cap):
   0.02% proprietary reagent for controlled ferrofluid aggregation, 0.5% BSA, and 0.1% sodium azide in buffer.
5. 3.0 mL Permeabilization Reagent (green cap):
   0.011% proprietary permeabilization reagent and 0.1% sodium azide in buffer.
6. 3.0 mL Cell Fixative (red cap):
   25% proprietary fixative ingredients, 0.1% BSA, and 0.1% sodium azide in buffer fixes cells for identification and enumeration.
7. 2 × 110 mL Dilution Buffer:
   A buffer with 0.1% sodium azide is used for dilution. It is the only reagent of the kit, which should be *stored at room temperature after opening the first time* (*see* **Note 2**).
8. 16 CellSearch®Conical Centrifuge Tubes (15 mL) and Conical Tube Caps.
9. 16 CellSearch® Cartridges and Cartridge Plugs.

2.1.4 CellSearch® Circulating Tumor Cell Control Kit

The CellSearch® Circulating Tumor Cell Control Kit is used for testing system performance and operator technique. It is sufficient for 24 applications and proves that the system detects both high as well as low numbers of CTCs. It has to be stored at 2–8 °C and contains:

1. 24 single-use vials of fixed cells from breast carcinoma cell line (*see* **Note 4**).
2. 24 lot-specific barcodes (*see* **Note 5**).

2.1.5 Additional Materials

1. CellSave Preservative Tubes for collecting 7.5–10 mL of sample (*see* **Note 6**).
2. CELLTRACKS® AUTOPREP® Instrument Buffer for the enrichment and staining procedure.
3. De-ionized water.
4. 0.26% sodium hypochlorite cleaning solution (*see* **Note 7**).
5. Horizontal swing out style rotor (i.e., swing bucket) centrifuge capable of 800 × *g* and break of "0."
6. Test tube racks.
7. Calibrated micro-pipettors and tips.
8. Vortex mixer.

2.2 Sample Preparation* for (Single)-Cell Isolation with the DEPArray™ Platform

1. CellSearch® Cartridge with enriched and stained cell suspension.
2. BSA lyophilized powder, essentially globulin and protease free, $\geq$ 98%.
3. 1 ×PBS.
4. Buffer for the DEPArray™ system (DABUF).
5. Protein gel loading tips round convenient for 200 μL-micropipettes.
6. 1.5 mL Protein LoBind tubes.
7. 50 mL screw cap tubes.
8. Racks for 1.5 mL and 50 mL tubes.
9. Micropipettes (10, 20, 200, 1000 μL) and related LoRetention filter tips (*see* **Note 8**).
10. Swinging bucket rotor centrifuge with adapters (*see* **Note 9**) for 1.5 mL tubes.
11. Laminar flow hood.

    *Alternatively, the VR volume reduction device (Menarini Silicon Biosystems) can be used following the instructions of the user manual to reduce the volume prior to loading the DEPArray™ cartridge.

2.3 (Single)-Cell Isolation with the DEPArray™ Platform

1. DEPArray™ Plus for CTC Application Guide and DEPArray™ Instruction.
2. DEPArray™ system (*see* Fig. 3).

Fig. 3 DEPArray™ Plus System. (© by Menarini Silicon Biosystems SpA)

3. DEPArray™ cartridges.
4. Buffer for the DEPArray™ system (DABUF).
5. 1.5 mL Protein LoBind tubes.
6. MicroAmp Reaction Tubes with cap 0.2 mL.
7. Micropipettes 5000 μL, 20 μL and related LoRetention filter tips (*see* **Note 8**).
8. Ultrasonic bath with a pulse button.
9. Racks for 1.5 mL and 0.2 mL tubes.
10. Laminar flow hood.
11. Waterproof pen for the unique labeling of the tubes.

2.4 Volume Reduction of DEPArray™ Isolated Cells*

1. 0.2 mL MicroAmp Reaction tubes with the DEPArray™ isolated cells inside.
2. 1 × PBS.
3. 200 μL micropipette and related LoRetention filter tips (*see* **Note 8**).
4. Fixed rotor centrifuge with adapters for 0.2 mL tubes.
5. Racks for 0.2 mL tubes.
6. Freezer for storage.

*Alternatively, the VR volume reduction device (Menarini Silicon Biosystems) can be used following the instructions of the user manual to reduce the volume after the isolation of the cells.

3 Methods

3.1 CTC Enrichment with the CELLTRACKS® AUTOPREP® System

Before starting a run, the system must be cleaned (*see* **Note 10**). Eight samples can be processed in parallel. Make sure that the samples fulfill all requirements for the run (*see* **Note 11**). The necessity for a control sample arises whenever patient samples are processed or when using a new lot number of kit reagents (*see* **Note 12**). Make sure to wear the required protective clothes and gloves for all following steps.

1. Start the CellTracks® Autoprep® System.
2. Bring the CellSearch® Circulating Tumor Cell Kit and one bottle of the CellSearch® Circulating Tumor Cell Control Kit to room temperature (*see* **Note 1**).
3. Select "Run batch" and enter the password (*see* **Note 13**).
4. Make sure, that the system is clean. If necessary, perform the "Daily Cleaning Procedure" (*see* **Note 10**), otherwise go on with **step 4**.
5. Take as many CellSearch®Conical Centrifuge Tubes and Caps as there are samples (including positive control if necessary). Put a barcode label (if available, *see* **Note 14**) on the CellSearch®Conical Centrifuge Tubes for the samples (*see* **Note 15**).
6. Mix the CellSave tube by inverting it five times (*see* **Note 16**).
7. Remove the cap of the CellSave tube and transfer 7.5 mL of the blood sample into a CellSearch® Conical Centrifuge Tube. Add 6.5 mL dilution buffer, close it with a cap and mix gently by inverting five times (*see* **Note 16**).
8. Repeat **steps 6** and **7** for further samples.
9. Centrifuge the tubes with a horizontal swing-out style rotor at 800 g for 10 min at room temperature. *Attention*: Set break to "0" (*see* **Note 17**).
10. Prepare the positive control—if necessary—during the centrifugation step, otherwise switch to **step 13**.
11. Put an orange ID label onto the conical tube with the positive control (*see* **Note 5**).
12. Mix the 3 mL control tube gently by vortexing 5 s and inverting the bottle five times (*see* **Note 18**). Open the cap and transfer the complete volume—also from the cap—to a conical centrifuge tube (*see* **Notes 19** and **20**).
13. Open all bottles of the CellSearch® Circulating Tumor Cell Kit and inspect them (*see* **Note 21**).
14. Follow instructions on the screen for choosing the kit, selecting a marker and control.
15. After choosing the number of samples, including the control, the system is "homing," which means being prepared for the run.

16. Place the rack with the CellSearch® Circulating Tumor Cell Kit into the CellTracks®AutoPrep® System (*see* **Note 22**) and confirm the kit data with “Next.”
17. Confirm, if results shall be used for patient management (*see* **Note 23**) or not. (If the kit is not valid anymore for in vitro diagnostics, *see* **Note 3**).
18. Put an empty CellSearch® cartridge (without lid) into the magnetic rack (MAGNEST®) and load it to the system (*see* **Notes 24** and **25**).
19. Repeat **step 18** for all remaining MAGNEST®s.
20. Go on with loading the samples and/or control (*see* **Note 26**). If a control is included in the run, follow the loading order in analogy to the order of the MAGNEST®s which were loaded in **step 18** (*see* **Note 24**).
21. Close the door and start the system with “Start.” During the run, the estimated duration is indicated (*see* **Note 27**).
22. When the enrichment is finished, remove the MAGNEST®s from the instrument. Carefully inspect the content of the CellSearch® cartridge (*see* **Note 28**).
23. Store the MAGNEST®s with the filled and closed CellSearch® cartridge at room temperature at a dark place in a horizontal position and analyze them with the CELLTRACKS ANALYZER II® system within the next 24 h (*see* **Note 29**). This is described in Subheading 3.2.
24. Follow the instructions on the screen for removing the empty sample/control tubes and inspect them (*see* **Note 30**). Remove the kit and close the reagents (*see* **Note 31**).
25. Go on with the daily cleaning procedure if no further runs are planned.

3.2 CTC Counting with the CELLTRACKS ANALYZER II® System

Keep in mind that enriched samples (also controls) can be scanned at the earliest after 20 min of incubation (in a horizontal position in the dark at room temperature).

1. Switch on the device, computer, and the mercury vapor lamp (*see* **Note 32**). The CELLTRACKS® software starts automatically.
2. Sign in with the unique password.
3. If it is necessary to perform a system verification (*see* **Note 33**), click the “QC test” folder symbol, clean the system verification cartridge with a fiber-free wipe, open the sample door, and load it to the sample rack. Push the “Start” button in the field “System verification” (*see* **Note 34**). After successful verification reload the verification cartridge.

4. Check if there are ferrofluid aggregates in the first CellSearch® sample cartridge, but leave the cartridge in the MAGNEST® (*see* **Note 28**).
5. Scan enriched patient samples by clicking to the "Patient Test" folder. For scanning the CellSearch® control cartridge go to **step 23**.
6. Clean the cover of the CellSearch® sample cartridge with a fiber-free wipe and load the MAGNEST® with the sample to the sample rack.
7. Check if the sample and kit information are correct (*see* **Note 35**).
8. Start scanning. When the edges of the CellSearch® cartridge are found, confirm by clicking "Accept" (*see* **Note 36**).
9. The system will start to focus automatically (*see* **Note 37**).
10. When focusing is finished the picture recording starts automatically (*see* **Note 38**).
11. After the scan is finished, click "Ok" (*see* **Note 39**), open the door, and remove the MAGNEST® with the CellSearch® cartridge (*see* **Note 40**).
12. Do not repeat scanning of a CellSearch® sample cartridge more than once (*see* **Note 29**).
13. Scan all other CellSearch® sample cartridges by repeating **steps 4–12** or go on with their evaluation (**step 14**). For analyzing CellSearch® control cartridges go to **step 25**.
14. For each sample scan, the system creates an auto-analysis of the pictures and shows an image gallery of cells, which have to be controlled by the user (*see* **Note 41**).
15. For evaluation, select the patient sample of interest in the software (in "Patient data" → "Sample table") and review the cells under the official guidelines of cell interpretation (*see* **Note 42** and refer to the user manual).
16. The auto-analysis generates an image of each event with every filter and defined filter combinations (*see* **Note 43**).
17. Select each event in the DAPI/CK-PE channel (*see* **Note 44**), which fulfills the criteria for CTCs (*see* **Note 42** and user manual). The image is now framed orange and the system counts this event as a CTC.
18. Make sure that all images of the gallery have been reviewed, otherwise evaluation cannot be finished (*see* **Note 45**).
19. The status of the sample changes to "Complete." The number of assigned and unassigned cells is shown.

20. If the quality of the scan is insufficient, the result is denoted "Done—No Result."
21. For printing or saving the data, click "Patient Data" → "Sample Table," select a sample, and click "Report". When the review of this sample is finished, a pdf document can be printed or exported.
22. Save the data from time to time to a separate storage location, e.g., by burning a DVD (*see* **Notes 46** and **47**).
23. Scan the enriched control by clicking "QC test."
24. Proceed with **steps 6–11** and treat the CellSearch® control cartridge just like the CellSearch® sample cartridges.
25. The system creates an auto-analysis with a pre-count of most of the "high" control cells. Nevertheless, the evaluation of some of the "high" and all of the "low" control cells (shown automatically at "QC Data") has to be finished manually by the user to classify the remaining, not automatically dedicated cells either to the high or the low row (*see* **Note 48**).
26. If the number of the "high" and "low" cells are in the correct range (*see* **Note 4**), the in vitro diagnostic (IVD) control has "passed."
27. The CellSearch® control cartridge can be discarded and the system switched off.
28. If there is no sample or QC to report, the computer can be shut down.

3.3 Sample Preparation for Single-Cell Isolation with the DEPArray™ System

After enrichment of CTCs with the CellSearch® system, the sample needs to be subjected to a buffer change before applying it to the DEPArray™ cartridge (*see* Figs. 4 and 6) for automatic isolation of single CTCs or blood leukocytes or pools thereof. All following working steps should be performed under a laminar flow hood wearing the required protective clothes and gloves.

1. Thaw a vial of DABUF.
2. Prepare a 2% BSA in PBS solution (*see* **Note 49**).
3. Prepare.
   - two aliquots of 325 μL DABUF.
   - one labeled 1.5 mL Protein LoBind tube for the sample.
   - 1.5 mL 2% BSA/PBS solution (*see* **step 2**).
   - one clean 1.5 mL tube for storing the tip.

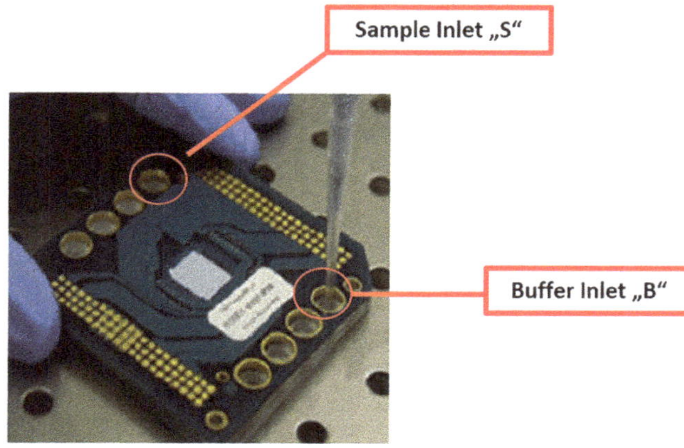

Fig. 4 Backside of the DEPArray™ cartridge with buffer and sample inlet. Picture adapted from the DepArray™ User Manual. (© by Menarini Silicon Biosystems SpA)

4. Remove the plug of the CellSearch® cartridge (*see* **Note 50**) which contains the enriched and stained cell suspension.
5. Set the micropipette to 200 μL and coat one protein gel loading tip with the 2% BSA/PBS solution (*see* **Note 51**).
6. Use this tip to resuspend the sample by pipetting the sample against all sides and the bottom of the CellSearch® cartridge. Transfer first 200 μL and then the rest (~125 μL) of the sample to the prepared and labeled 1.5 mL Protein LoBind tube and keep this tip for later use (*see* **Note 52**).
7. Use a new protein gel-loading tip for transferring the first aliquot of the 325 μL of DABUF into the empty Cell-Search® cartridge in two steps. Throw away this tip.
8. Use the pre-coated tip from **step 6** for washing the Cell-Search® cartridge with the DABUF by pipetting the buffer against all sides and the bottom. Transfer the suspension in two steps into the 1.5 mL tube, which already contains sample (**step 6**) without resuspending. Keep this tip for later use (*see* **Note 52**).
9. Repeat **steps** 7 and **8** with the second 325 μL aliquot of DABUF.
10. Centrifuge the labeled sample tube with a swinging bucket rotor and the 1.5 mL tube adapters (*see* **Note 9**) with 1000 g for 5 min at room temperature.
11. Put the tube into the rack under the laminar flow hood, open it carefully to avoid swirling up the cells, and remove the supernatant (*see* **Note 53**) without touching the

bottom of the tube with the 1000 μL micropipette. Leave a volume of approx. 30 μL in the tube.

12. Add 1 mL DABUF *without resuspending*. Close the tube and flip with the finger against the wall to wash the pellet gently.
13. Repeat **step 10**.
14. *Carefully open the tube under the laminar flow hood. Remove the supernatant (*see* **Note 53**) without touching the bottom of the tube with the 1000 μL micropipette, then with the 200 μL micropipette, and finally with the 20 μL micropipette (*see* **Note 54**) and leave around 10 μL in the tube.
15. Adjust the volume to 12 μL (*see* **Note 55**) without touching the cell pellet. Do not throw away the 20 μL tip (compare its storage with **Note 52**), keep it for loading the DEPArray™ cartridge.
16. Go on directly with loading the DEPArray™ cartridge (*see* Subheading 3.4).

    *As an alternative for manual volume reduction (**steps 14** and **15**), the VR volume reduction device (Menarini Silicon Biosystems) can be used following the instructions of the user manual to reduce the volume prior to loading the DEPArray™ cartridge.

3.4 Cell Isolation with the DEPArray™ System

All steps concerning the loading of the DEPArray™ cartridge have to be performed under a laminar flow hood wearing the required protective clothes and gloves.

1. Take the DABUF bottle and incubate them for 10 min in the ultrasonic bath with the degas function (*see* **Note 56**).
2. Start the DEPArray™ system. Select the username in the dropdown menu to sign in. Enter the password by touching the "Code" button (*see* **Note 57**) and login.
3. Enable the VPN connection in order to have full service for troubleshooting.
4. Touch the "Sorting," "CTC," "UDP fixed CTC" buttons on the touch screen (*see* **Note 58**).
5. Go back to the laminar flow hood with this DEPArray™ cartridge. Open cover "1" (*see* **Notes 59** and **60**), remove the blue protective tape, and leave the DEPArray™ cartridge inside the laminar flow hood for loading.
6. Pipette 2.5 mL of the degassed and solved buffer (**step 1**) to the buffer reservoir "B" (*see* Fig. 4) of the DEPArray™ cartridge by using the 5000 μL micropipette (*see* **Note 61**).

7. Check at the back of the DEPArray™ cartridge, if there are air bubbles inside the buffer channel. Also make sure that the meniscus of the buffer has not entered the valve of the main chamber (*see* **Note 62**).
8. If there are no bubbles in the buffer channel, go on with pipetting (*see* **Note 63**) the 12 μL sample (prepared in Subheading 3.3) into the sample channel "S" (*see* Fig. 4) of the DEPArray™ cartridge.
9. Check at the back of the DEPArray™ cartridge, if there are air bubbles inside the sample channel and make sure that the meniscus is in the correct area (*see* Note **64**). Close the protective cover of the DEPArray™ cartridge and go back to the DEPArray™ system for loading the DEPArray™ cartridge.
10. For opening the door of the system, touch "Open" (*see* **Note 65**).
11. Open cover "2" (*see* **Note 59**) of the DEPArray™ cartridge protection cover and insert DEPArray™ cartridge into the correct position of the tray (*see* **Note 66**).
12. Close the door of the system by pushing it gently. The cartridge tray is automatically pulled back, the door is locked, and the system recognizes the cartridge ID.
13. Scan the barcode of the buffer. This step can also be skipped by touching "Skip."
14. Fill in the run code (name of experiment) by touching "Run Code" and using the keyboard on the touchscreen of the machine.
15. Touch "Start Process" to start the sorting execution.
16. The sample loading phase (*see* **Note 67**) is starting automatically.
17. After the sample loading is finished, the sample scan (*see* **Note 68**) starts automatically.
18. For optimizing the scan settings, push the "Pause" field on the right side of the monitor as early as possible to pause the scan.
19. Click several frames (*see* **Note 69**) in the DAPI channel for finding cells and check the filter setting parameters for all fluorescent channels (*see* **Note 70**). Once adapted correctly push the "Play" button again. The system asks for confirmation of the changes in chip scan settings. By clicking "Yes" the system restarts scanning all frames under the new conditions. If no re-scan shall be done under the new conditions, click "No" and the system will go on with scanning from the position where it had stopped before. The scan of the cartridge takes around 30 min.

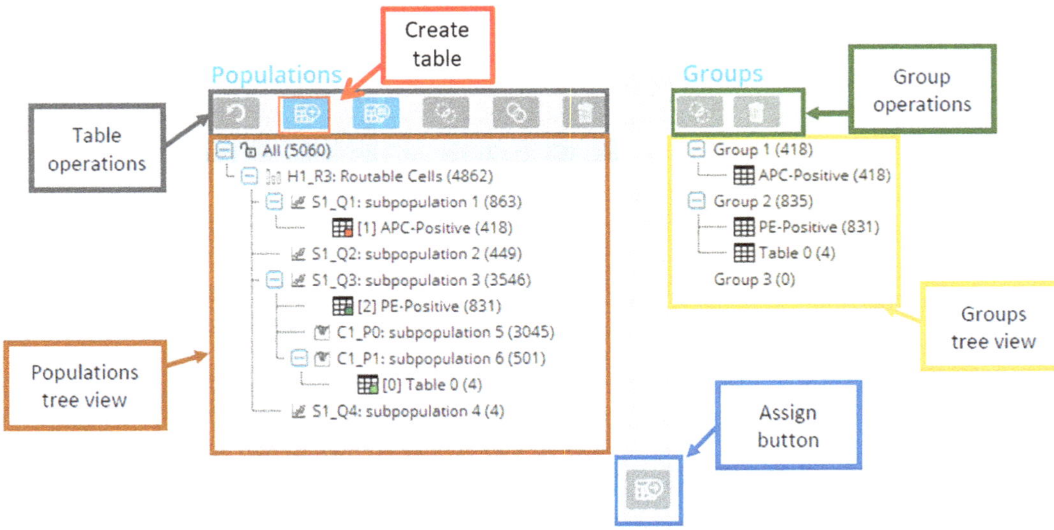

Fig. 5 Populations and Groups panels of the CellBrowser™ with functional buttons. Picture adapted from the DepArray™ User Manual. (© by Menarini Silicon Biosystems SpA)

20. When the scan is finished, the cell selection workflow can be started using the "Cell Selection Software" (*see* **Note 71**).
21. An automatic selection of cells (*see* **Note 72**) is shown in the "Populations" panel (*see* Fig. 5).
22. For an overview of the cell list and the plot view in parallel, click the "Grid and Data Visualization" icon.
23. Select the "Routable Cells" (*see* **Note 73**) by clicking on it (they will be highlighted in blue) and click on the "Scatter Plot" icon in the "Data Visualization" panel.
24. Select "Mean Intensity PE" for X-axis and "Mean intensity APC" for Y-axis. Click the "Cross Selection" button to split the "Routable Cells" selected for PE and APC into the four subpopulations and move the lines for optimization (*see* **Note 74**).
25. Select the PE^+/APC^- subpopulation ("Q3") and click the "Create Table" icon (*see* Fig. 5) above the cell populations in the "Populations" panel. A new table, called "[0]Table0 (0)," is created (*see* **Note 75**) and will be filled in with the cells of interest to be isolated.
26. Sort the cells in the "Grid View" list from highest to lowest mean intensity of PE by clicking twice in the blue "Signal Intensity PE" field (*see* **Note 76**) on the top of the list.
27. By clicking to the first cell in the "Grid View" list (highlighted in blue), the fluorescent images of the selected cell are shown below. Judge the cell by the official cell selection parameters (*see*

Note 77). Shift putative CTCs from the Q3 subpopulation to the table "[0]Table0 (0)" (*see* **Note 78**) by typing the number in squared brackets, i.e., 0 for "[0]." Proceed with all cells in the "Grid View" and shift the putative CTCs to the respective table.

28. Once all putative CTCs are selected, analogously repeat **steps 25–27** for the selection of the putative leukocytes (APC^{+}/PE^{-} subpopulation = Q1, create a new table for this subpopulation) or other cell populations if needed.
29. In order to route the selected cells, the tables must be assigned to groups in the "Groups" panel. For this, first highlight the table containing the cells selected for routing, e.g., "[0]Table0 (XY)" in the "Populations" panel and then "Group1" in the "Group" panel. For correlating both click the blue "Assign" icon (*see* Fig. 5) between the "Populations" and the "Group" panel. Repeat this for all tables containing cells to route (*see* **Note 79**). The groups are automatically saved in the "Cell Panel" (*see* **Note 80**).
30. Click the "Cell Routing" icon (top right) at the CellBrowser™ to start the "Recovery Manager™" (*see* **Note 81**).
31. Open the shutter in the "Live Configuration" section on the lower left side. The DEPArray™ cartridge chamber becomes visible. The squared frame in the main chamber view (top right) can be moved for checking, if air bubbles are inside the DEPArray™ cartridge chamber (*see* **Note 82**).
32. Highlight the first group to be routed to the "Park" section by left-click. The group is marked to be parked in the parking chamber by right-clicking and selection of "Move To Routing." Repeat this for all groups to be parked and start "Park Routing" by clicking the play button (*see* **Note 83**).
33. When the "Park Routing" is done a green check and the comment "Park done" in the "Park" section notify the user about successful parking (*see* **Note 84**).
34. For defining the recovery scheme, select the appropriate "Recovery Support" in the software. For the "CTC-RUO Fixed" application, the "200 μL-tube-rack" is preset and recommended for maximum purity of the samples (*see* **Note 85**).
35. Select position "A1" in the "Recovery Support" layout by right-click and set it as the priming recovery position (*see* **Note 86**). To recover the putative CTCs as cell groups (in order to obtain pools of cells), click the cell group name. To recover single cells, click the cell ID. Drag selected group or cell ID to the selected position of the "Recovery Support" layout (*see* **Note 87**).

36. If more than one cell type is to be recovered (e.g., putative CTCs and putative leukocytes), a blank recovery in between each cell population should be defined (*see* **Note 88**).
37. To recover further cell types (e.g., putative leukocytes) as cell groups or single cells, proceed in analogy to the CTC recovery using the drag and drop function (*see* **Note 87**).
38. Prepare the necessary number of clean 200 μL tubes (*see* **Note 89**), label them with a unique name (corresponding to the experiment), and put each tube to the respective position selected in step 37/39 (*see* **Note 90**).
39. For insertion of the loaded "Recovery Support" into the machine, click the "Play" button in the "Exit and Recovery" section. Select the appropriate support "200 μL-tube-rack" (or the support chosen in **step 38**, *see* also **Note 85**) in the drop-down menu and click "Add."
40. Switch to the DEPArray™ touch screen and touch "Trays" and "Open Recovery" to eject the recovery tray.
41. Open the main door and insert the "Recovery Support" with the opened 200 μL tubes (labeled) in the correct position (consider the marks on the tray) and close the main door of the DEPArray™ system.
42. Touch "Close Recovery" and switch to the recovery software on the screen. A check box icon is shown, when the "Recovery Support" is inserted correctly and has to be confirmed with the "Ok" button.
43. After confirmation of the pop-up window, the initial washing step starts (*see* **Note 91**). After the washing step, the system automatically starts with the "Priming Recovery" and the recovery of the cells (*see* Fig. 6).

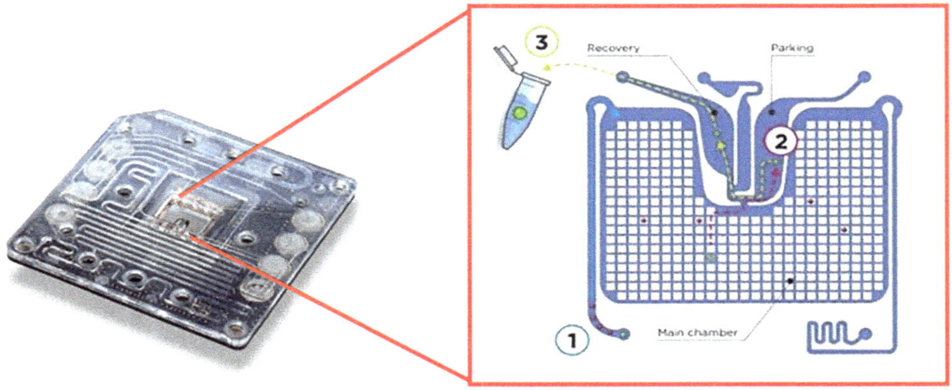

Fig. 6 Front side of the DEPArray™ cartridge with the three-chamber system as the heart of the cartridge. (© by Menarini Silicon Biosystems SpA)

44. When the recovery of a cell group or a single cell is completed, a green check appears in the "Exit and Recover" field.
45. When recovery of all cells is finished, the samples can be removed. Push the "Stop" button on the touch screen of the DEPArray™ system and confirm with "Ok."
46. Sign in with username and password (*see* **Note 92**).
47. Touch "Trays" at the touch screen of the DEPArray™ to open the main door for removing the "Recovery Support" with the samples and check if the tubes are filled.
48. Close the caps of the 200 μL PCR tubes.
49. Close the main door by touching "Close Recovery" on the touch screen of the device.
50. Touch "Eject Cartridge" to open the DEPArray™ cartridge tray and remove and throw away the used DEPArray™ cartridge (*see* **Note 93**). Close and lock the main door by touching "Insert Cartridge."
51. Touch "Back," "Shutdown," and "Ok" for shutting down the device.
52. Before the system shuts completely down, a backup is performed automatically (*see* **Note 94**). Go on with **step 53** if no further experiment is planned for this day or stop the backup process, if another experiment follows (*see* **Note 95**).
53. Switch off the machine at the end of the backup process (when the touchscreen is black and does not respond by touching it).

3.5 Preparation of Isolated Cells for Molecular Analyses

Single cells or cell pools are recovered from the DEPArray™ system in a cell number-specific volume (*see* **Note 96**). In order to proceed with whole genome amplification (WGA) followed by molecular analyses, the volume of the recovered single cells or cell pools needs to be reduced to approx. 1 μL. Volume reduction of the isolated samples has to be performed under a laminar flow hood wearing the required protective clothes and gloves.

1. Spin down the 200 μL tubes at room temperature and 14,100 g at room temperature in a fixed rotor centrifuge for 200 μL tubes. The time depends on the number of cells in the tubes (*see* **Note 97**).
2. Proceed with the following steps under a laminar flow hood.
3. If 21–85 cells are recovered in one tube (compare to **Note 96**), carefully remove 50 μL of supernatant. Do not touch the bottom of the tube. Instead, follow the meniscus of the solution.
4. Add 100 μL of 1× PBS to all tubes (regardless of prior 50 μL removal of the supernatant) without touching the solution.

5. *Spin down the samples at room temperature with 14,100 g for 10 min at room temperature in a fixed rotor centrifuge and consider the tube cap direction (compare to **Note 97**).
6. For reducing the volume to approx. 1 μL, go back under the laminar flow hood, open the tube in the rack, and hold it horizontally.
7. Be sure to stabilize arms by resting the elbows on the bottom of the hood.
8. Aspirate the supernatant carefully and slowly by following the meniscus and not touching the bottom of the tube until approx. 1 μL is left. Use a 200 μL micropipette which is set to 150 μL. Use a fresh tip for each tube (*see* **Note 98**).
9. The sample is now ready for whole genome/transcriptome amplification and subsequent molecular analyses.
10. Alternatively, the sample can be stored at −20 °C after volume reduction.

    *As an alternative for manual volume reduction (**steps 5–8**), the VR volume reduction device (Menarini Silicon Biosystems) can be used following the instructions of the user manual to reduce the volume after the isolation of the cells.

4 Notes

1. The kit has to reach room temperature before it is used. Ideally, the kit is transferred to room temperature approx. 30 min before sample processing is started. To avoid evaporation, remove the caps just directly before the run starts.
2. Once the dilution buffer has reached room temperature at first opening, further storing has to be done at room temperature and the buffer is stable for 4 weeks. Cooling of the buffer can influence the sample and associated enrichment quality. The dilution buffer is part of the CellSearch® Circulating Tumor Cell Kit.
3. The CellTracks® Autoprep® System saves the first opening date of the kit by scanning the kit barcode. To guarantee stability under FDA guidelines, the kit can be used only 30 days for patient management, thus diagnostic purposes. For research/nondiagnostic experiments, it is possible to use the kit until the date of expiry. The system has a safety warning message installed, which precludes enrichment of important diagnostic samples with a kit open for more than 30 days.
4. Each vial of the CellSearch® Circulating Tumor Cell Control Kit contains two differently labelled cell types. A high number of CK-PE$^+$/DAPI$^+$/FITC$^+$/APC$^-$ cells (mean ~ 1000 cells;

"high" cells) is detected by the FITC-channel and a low number of CK-PE$^+$/DAPI$^+$/APC$^+$/FITC$^-$ cells (mean ~ 50 cells; "low" cells) is detected by the APC-channel. The exact range of the cell numbers is indicated on the label of the control kit box and can vary between different lot numbers. The number of cells, detected in the control cartridge should fall into this range, otherwise the control is regarded to have failed.

5. The orange barcode label included in the control kit contains the lot-specific information (lot number, date of expiry, mean cell number of "high," and "low" cell type) and informs the system, to skip the plasma aspiration, as it is done for the patient samples.
6. Once the blood is collected in the CellSave Preservative Tube, the proprietary preservative stabilizes the cells for up to 96 h. Store the blood samples at room temperature until the run starts and inspect the sample for clots before it is processed. Wrong storage of the sample can have an influence on the quality of the enrichment with the sensitive ferroparticles.
7. The bleacher Clorox® is suggested for the cleaning procedure. Alternatively, it is also possible to prepare a 0.26% sodium hypochlorite solution in a final volume of 4 L.
8. It is possible to use other pipettes with LoRetention filter tips than the suggested ones from Eppendorf. However, according to our experience, this combination fulfills the best requirements for an uncomplicated bubble-free run. Alternative tips can be used as suggested from Menarini Silicon Biosystems.
9. The adapters for the 1.5 mL tubes are 50 mL screw cap tubes with a hole, centered in the screw caps, fitting for 1.5 mL tubes. They can also be self-made, if the balance of the centrifuge is guaranteed.
10. The daily cleaning procedure should be performed:
    (a) at the end of each day after sample processing.
    (b) at least every 72 h.
    (c) if a sample with more than 5000 CTCs was included in the last batch.

    Start the system, choose "Daily cleaning," and enter the password. Follow the instructions of the system for disconnecting the bottle with the AutoPrep® Instrument buffer and for connecting the bottles with de-ionized water, cleaning solution and waste. Push "Enter" and the cleaning procedure starts. It takes around 1 h. The system shows a reminder every 72 h to perform the daily cleaning procedure. Follow the instructions after finishing the cleaning procedure and push "Enter" to go back to the main menu. For more information about cleaning, weekly and monthly maintenance, please refer to the user manual.

11. Only process samples which meet the following requirements:
    (a) The sample was stored at room temperature in the Cell-Save tube no longer than 96 h.
    (b) The sample has volume of at least 7.5 mL.
    (c) The sample is not coagulated.
12. Not every run has to include a control sample. But it is necessary to run a control sample.
    (a) once a day when patient samples are processed.
    (b) when a reagent kit with a new lot number is used.

    Keep in mind, that only seven more patient samples can be processed at a time if a control is included.

    If the system realizes a new lot number of the CellSearch® Circulating Tumor Cell Control Kit, the barcode of the package has to be scanned to inform the system about the lot-specific informations (*see* **Note 5**). It is possible to run more than one control in one run.
13. Make sure that the de-ionized water bottle is disconnected, the bottle with AutoPrep® Instrument buffer is filled and connected, the waste bottle had been emptied, filled with 400 mL undiluted sodium hypochlorite solution and connected.
14. There are no barcode labels included in the kit for the samples. If it is necessary to code the sample tubes digitally, the CellTracks® AutoPrep® System can decode barcodes with the CLSI-Standard Auto02-A2. The label should be no longer than 50 mm. Fix the barcode sticker in the upper part of the conical tube above the sample level. This ensures that the blood sample or the control solution are not covered.
15. If there are no barcode stickers available, mark the sample tube above the sample level with a unique sample name. At a later stage of processing, it is possible to enter the sample name to the system (*see* **Note 26**).
16. Invert the tube in order to obtain a homogeneous sample. This is important for further processing.
17. The time to reach 800 g before and 0 g after centrifugation is not included in the 10 min of centrifugation. Since the break is set to "0," it will take some minutes longer until the centrifuge has finished.
18. For an improved and more homogeneous suspension of the control cells, it is advisable to open the bottle and pipette five times up and down with the 1000 μL micropipette and a clean filter tip.

19. Tip over the empty and open control bottle and collect remaining drops in the CellSearch®Conical Centrifuge tube for the control.
20. There is no need for adding dilution buffer and centrifuging the control tube.
21. Each bottle of the tray should be fixed properly by pushing them down. To avoid evaporation, remove the caps directly before the run. An inspection of the solutions in the bottles for air bubbles is necessary before each run. A clean micropipette tip can be used to remove them.
22. This is the last possibility to ensure that the reagents are at room temperature. The system reports if the number of samples to be processed does not match the number of remaining samples which can be processed with this kit.
23. Samples which are used for the patient management are always written in black, on each and every CELLTRACKS®AUTOPREP® System.
24. Every stained and enriched sample is collected in a single-use CellSearch® cartridge, which is entered in a magnetic rack, the so called MAGNEST®. There is no rule specifying the order of loading samples and controls. However, in order to prevent confusion, the gray MAGNEST® should be used for the control and the purple MAGNEST®s for the samples. Make sure to load control/samples and gray/purple MAGNEST®s accordingly using the same order. Every MAGNEST® has a memory chip saving all the information created during the run. It is deleted before every new run automatically, so the MAGNEST® can be used continuously.
25. Each CellSearch® cartridge should be empty and clean inside. The outside can be cleaned with a fiber-free tissue. The appropriate number of CellSearch® cartridges (corresponding to the number of controls/samples) are inserted into the MAGNEST®s and fixed properly by pushing them down. MAGNEST®s must be undamaged since this could influence the recovery of cells and the associated report. After entering a MAGNEST®, the system shows the barcode of the CellSearch® cartridge on the screen, closes the door, and moves the MAGNEST® holder carousel to the next free position. All filled MAGNEST®s have to be inserted.
26. The samples/controls have to be loaded in a way that the barcode is directed toward the user and the sample tube is freely rotatable. If there is no barcode label on the sample tubes, the sample can be loaded anyway. By pushing "Enter," a warning message appears (*see* also **Note 15**) and the sample ID and type can be edited. This information has to be saved.

27. The run takes 2 h 20 min plus 13 min per additional sample.
28. The CellSearch® cartridge must not be removed from the MAGNEST® for inspection! It should be filled with an orange-yellow colored solution (if not *see* **Note 30**). Be aware that in rare cases, ferrofluid reagent can aggregate because of disturbing factors in the blood of some patients. As a consequence, the ferrofluids might mask CTCs during the scan which could influence the results. Aggregated ferrofluids close to the plug might have less influence to the results than aggregates in the middle of the sample. If there are bubbles inside the CellSearch® cartridge, place the plug onto the CellSearch® cartridge opening without pushing it down. While keeping the plug in its place, knock the MAGNEST® gently against a hard surface (e.g., bench) to destroy the bubbles. Then re-seal the CellSearch® cartridge with the plug. If the CellSearch® cartridge is empty, the associated conical tube might still be filled with liquid. If this is the case, please seek advice from the technical support.
29. The closed MAGNEST®s holding the CellSearch® cartridge should be stored for a minimum of 20 min before scanning in a horizontal position. This ensures that ferrofluid connected cells sediment to the same layer for scanning. IVD samples and controls have to be scanned with the CELLTRACKS ANALYZER II® after 24 h at the latest and are allowed to be scanned only twice. Note that if samples are scanned more than twice within 24 h, they lose their IVD status.
30. Every CellSearch®Conical Centrifuge tube has to be inspected before throwing it away. Ferrofluid accumulations might have remained inside the tube. The results of such samples should be considered with care since they might possibly be invalid.
31. Before removing the kit, it is possible to make a print-out with the information of how many runs can still be performed with the opened kit. For this, go back to the main menu ("View Data" → "Reagent Kit") and print. For removing the kits, close the bottles with their uniquely colored caps and store the kit until next use at 2–8 °C. The dilution buffer has to be stored at room temperature.
32. The mercury vapor lamp needs to warm up. This takes approx.15 min. The lamp is switched on either after confirming an automatically appearing request for warming up the lamp or by starting it manually by clicking the button "lamp" in the toolbar. The lamp signal blinks during warm-up phase and remains green when the lamp is ready. The lamp has a total run-time of 300 h. In order to save lifetime of the lamp, it is advisable to switch it off when there are no samples to be scanned and to avoid repeated switching on and off.

33. A system verification notice appears whenever the CELL-TRACKS ANALYZER II® had been turned off, the application was restarted, and 10 h have elapsed since the last system verification. It verifies the optical performance and the chamber skew. The system verification cartridge can be used for 180 verifications or until expiry date, depending on what comes first. If it is not in use, protect the system verification cartridge from light by storage in the corresponding box. Since the chip of the MAGNEST® records (and counts) the use of the system verification cartridge, the same MAGNEST® should always be used for the system verification cartridge.
34. If the system verification fails, a second trial or the use of a new system verification cartridge is possible. If it fails again, the user manual helps for focusing the crosshairs, filling in and saving the system verification information of the new system verification cartridge.
35. Adding comments or editing sample information should be done at this point. No changes can be done after the scan.
36. The system puts a rough focus to the ferrofluid and localizes the CellSearch® cartridge edges. When the edges are detected, four green dots appear on the image. If an edge is not detected, a red dot appears. In this case, the scan has to be stopped and it has to be confirmed that the CellSearch® cartridge has been inserted correctly into the MAGNEST®. If it still does not work, the edge has to be set manually by clicking to the undetected edge within the picture.
37. The automatic focusing is done on nine spots of the CellSearch® cartridge. Six of nine spots have to be successful for starting the scan.
38. The scan takes around 10 min. The CellSearch® cartridge is scanned with each filter frame by frame. Since the frames are overlapping, no cell is lost during the scan. After the scan is finished, the system compares overlapping regions automatically and removes duplicate cells.
39. Some frames appear gray after scanning. This means that the camera has optimized the settings for making faint objects visible.
40. The CellSearch® cartridge has to be removed from the MAGNEST® and stored in a vertical position at 4 °C in the dark until isolation of cells.
41. Results do not have to be evaluated immediately. All samples/controls can be scanned first.
42. Tumor cell phenotype:
    - CK-PE^{+}, the cytokeratine signal is positive (green in the composite image).

- DAPI$^+$, the nucleus signal is positive.
- CD45-APC$^-$, the specific signal for leucocytes is negative.
- The unlabeled channel is negative for tumor cells.

  Morphology:

  The object in the CK-PE channel should be an intact cell. This means:

- It has a size of at least 4 μM (compare to size of the cursor) and a round or oval shape (but also a polygonal or an elongated shape is possible).
- The nuclear area is smaller than the cytoplasmic area.
- More than 50% of the nucleus is visibly surrounded by cytoplasm.

43. When the gallery is shown the first time, the pictures are not assigned.
44. For a high-resolution picture of the event, the event number has to be clicked.
45. The evaluation does not have to be finished immediately. If the page is closed during evaluation, the sample state changes to "Review." This means that a review has been started. When each page of the gallery down to the last picture has been evaluated, the number of assigned and unassigned events is shown and the "Done" button is active for saving the sample evaluation. The sample status changes to "Complete."
46. Eight to twelve samples can be saved on a DVD + R. Only completely reviewed or "enabled" samples can be archived.
47. In the "Archive" folder, "Patient Samples" or "Control Samples" can be selected for archiving the results. By clicking "Archive" and selecting "Archive Oldest Results," the oldest results are stored to the DVD and confirmed with "Ok." If specific tests have to be archived together, the tests of interest have to be selected, confirmed by clicking "Ok" and selecting "Archive selected results." By clicking the "Start" button, the system automatically displays the number of tests to be burnt on one DVD and assigns a disc number. After labeling and insertion of a blank DVD, the archiving can be started with "Start."
48. Cells have to be labeled (an orange frame appears after selection) in the respective row where a definite signal (high or low) is seen. If one low and one high signal are in the picture, the event has to be counted as one low cell, and if two high cells are in the picture, it has to be counted as one high cell.
49. I.e., 0.5 g BSA in 25 mL 1 × PBS. The solution can be stored in 1.5 mL aliquots at 4 °C up to 1 week or at −20 °C up to 12 months. Avoid to freeze and thaw aliquots more than once.

50. By using a 1000 μL tip, the plug of the CellSearch® cartridge can be removed by levering on the backside of the plug. Alternatively, clean forceps can be used.
51. The pipette tip is coated by slowly pipetting BSA solution five times up and down. By immerging the tip down to the bottom of the tube, also the outside of the tip is covered. The coating of the tip prevents cells from sticking to it.
52. The used tip should be kept by putting it vertically in a clean open 1.5 mL tube placed in a 1.5 mL rack, the narrow end pointing up. To avoid contamination, the tip should not be touched at the narrow end.
53. The supernatants can be collected after the centrifugation steps in 1.5 mL tubes. This ensures that cells are not lost if the pellet had been disturbed and the isolation experiment had failed. In this case, the supernatants of the two washing steps should be combined, centrifuged for 5 min at 1000 g at RT, and reduced to 12 μL (*see* **Note 55**).
54. First approx. 800 μL with the 1000 μL micropipette, then approx. 150 μL with the 200 μL micropipette, and finally the rest of the solution with the 20 μL micropipette should be removed until approx. 10 μL are left. It is important not to touch the bottom of the tube, due to possible loss of cells. The 20 μL tip should be kept in the second clean 1.5 mL tube already prepared (as explained in **Note 52**) for adjusting the volume to 12 μL and loading the DEPArray™ cartridge.
55. The volume for loading the DEPArray™ cartridge has to be exactly 12 μL. To reach this, measure the volume left in the tube with the 20 μL micropipette and the stored 20 μL tip (from **Note 54**). After measuring, store the tip again in the 1.5 mL tube. If the volume is lower than 12 μL, add the missing volume with a fresh 20 μL tip (without touching the cell suspension on the bottom of the tube). If the measured volume is higher than 12 μL, centrifuge the sample using the adaptor (*see* **Note 9**) with 1000 g for 5 min at room temperature. Remove the extra volume with a 10 μL micropipette without touching the cell pellet for reaching exactly 12 μL of sample volume. Alternatively, the VRNxt device can be used for preparing the sample prior to loading.
56. The sonication of the buffer in the ultrasonic bath is done at room temperature and solves potential aggregations in the buffer. The activation of the degas function improves bubble-free buffer loading to the DEPArray™ cartridge.
57. It is possible to uncover the password by checking the box at "Show Code."

58. With this application, it is possible to park up to 105 single cells in the park chamber of the DEPArray™ cartridge and to recover single cells with high performance.
59. There are two possibilities to open the protective cover with the DEPArray™ cartridge inside. By opening cover "1," the DEPArray™ cartridge has the right position for being loaded (back side of the cartridge, see Fig. 4) with buffer and sample. By opening cover "2" (turn the protective cover upside down), the DEPArray™ cartridge is in the right position for being loaded to the system. In this position, the buffer and sample channels are visible as well as the main, the recovery, and the exit chambers forming the core of the cartridge (front side of the cartridge, see Fig. 6).
60. For loading the DEPArray™ cartridge, the instruction enclosed in the DEPArray™ cartridge package has to be followed. However, in order to correctly load the DEPArray™ cartridge and avoid both contamination of the system and sample as well as loss of important patient samples, it is strongly recommended to participate at a DEPArray™ user training. Among other applications, the user is trained the specific "CTC-RUO-Fixed Cells" application, which forms the basis for isolating CTCs with the DEPArray™ from a CellSearch® cartridge, enriched with the CellSearch® system.
61. The tip of the 5000 μL micropipette has to be filled without creating air bubbles. Those can influence the success of the experiment. Therefore, the loading tip has to be pressed firmly into the buffer inlet "B" and the pipetting speed should be very low.
62. In order to avoid air reaching the sensitive main chamber of the DEPArray™ cartridge, bulges to trap air bubbles are embedded in the buffer channel of the DEPArray™ cartridge. Bubbles collected in these "traps" or at the end of the channels usually do not cause problems during the run. In case of bubbles in the middle of the channels or at the meniscus close to the main chamber, the DEPArray™ cartridge is not usable anymore and needs to be exchanged for a new one.
63. For this step, it is recommended to use the 20 μL tip stored during Subheading 3.3 **step 15** in order to reduce the number of lost cells sticking to the pipette tip. Resuspend the sample gently without creating air bubbles before loading it on the DEPArray™ cartridge.
64. Once the sample meniscus reaches the main chamber during the loading step, the DEPArray™ cartridge cannot be used any more. After removing the sample from the DEPArray™cartridge, it can be reloaded to a new DEPArray™ cartridge (refer to user manual). However, this might cause a loss of cells. All

important steps for sample loading are described in the enclosed product insert of the DEPArray™ cartridge.

65. To avoid contamination, it is recommended to touch the touch screen of the device only with a pen.

66. For slotting in the DEPArray™ cartridge correctly, it is necessary to rotate it by 90° to the left while cover "2" is turned up. Open cover "2" and insert the DEPArray™ cartridge.

67. The "Sample Loading Phase" consists of a calibration step (an internal reference system in the chip is controlled) and the sample loading step (the recovery/exit chamber and the parking chamber are filled with DABUF and the main chamber is filled with the sample). The "Sample Loading Phase" takes around 15 min. The adjustment of the chip scan settings needs to be done at the end of the "Sample Loading Phase." Therefore, it is important to stay next to the machine in order not to miss the time point to pause the automatically starting "Chip Scan."

68. During this step, the surface of the main chamber is scanned with each filter selected during chip scan settings.

69. Due to the sample stream during loading, most of the cells locate to the lower right side of the main chamber. For checking the filter settings, it is therefore best to select frames in this part of the main chamber. However, selecting other frames is also possible.

70. All information for the correct adaption of the exposure time, the camera gain, and the offset focus can be found in the user manual. However, for a correct adaption of the filter settings, a training for the system is strongly recommended.

71. The CellBrowser™ and the "Cell Panel" are the two components of the "Cell Selection Software." The latter contains the "Grid View" (a list of detected cells), the "Data Visualization" (population plots), and the "Populations" (generation of subpopulations and selection groups) panel. Multiple options like histogram, scatter plot, position on the chip, set operator, and population cloning are available for informed cell selection (see user manual).

72. In the "Populations" panel, the detected events can be assigned to four hierarchical levels (*see* Fig. 5). The events detected automatically via image analysis and gating tools represent the input data and are called population "All" (= hierarchy one). The subpopulation "Routable Cells" (= hierarchy two) is also created automatically. (For events grouped to hierarchies three and four, *see* **Notes** 77 and **78**).

73. After an automatic calculation of routing ways between each cell and a parking position in the parking chamber, not all cells

may be routable. This may be due to the quality of the sample and the sample preparation procedure but also to particular algorithms for calculating the routes of the cells. The latter reflects a safety step of the system. In order to guarantee the purity of the routed cells, only those routes are considered which assure a free passage of the target cell without touching other cells or particles.

74. After the correct selection, the four subpopulation groups (APC^+/PE^- [leukocytes] = Q1, PE^+/APC^- [putative CTCs] = Q3, APC^+/PE^+ = Q2, PE^-/APC^- = Q4) are saved by clicking "Create" and appear as four groups in the hierarchical level three under the "Routable Cells" in the "Populations" panel (*see* Fig. 5). For a more detailed description of the "Scatter Plot Filtering" or other filtering options, follow the user manual.

75. For selecting cells which are to be isolated, a table has to be created. This is done by selecting the subpopulation (e.g., the PE^+/APC^- subpopulation = Q3) and clicking the "Create Table" icon. The new table called "[0]Table0 (0)" (= hierarchy four) appears under the subpopulation (in this case under Q3). This works in analogy for each subpopulation, e.g., for the APC^+/PE^- subpopulation (= Q1), a second table called "[1] Table1(0)" (= hierarchy four) is created, etc.

76. It is not obligatory to sort the cells in this way, but the CTCs are detected by the CK-PE signal and the signal strength correlates with the probability of the event representing a CTC. This is done analogously with the leukocytes (subpopulation Q1) by sorting the CD45-APC signal of this cell population from the highest to the lowest "Signal Intensity APC."

77. Several parameters are important for an optimal cell selection. Trapping, morphology, intensity, score, and other parameters (just available for some DEPArray™ applications) are described in detail in the user manual. To select the correct cells from the different subpopulations, it is possible to follow the user manual, but in order to avoid cross contamination of the populations of interest, an official user training is strongly recommended.

78. For "shifting," e.g., putative CTCs from the Q3 subpopulation to table "[0]Table0 (0)," the cell of interest in the "Grid View" list has to be marked and the table-associated number in squared brackets, in this case [0], has to be entered. The current number of cells appears in parentheses (e.g., "[0] Table0 **(1)**"). This can be applied to all following subpopulations and tables which have been created to select cells from. The number of cells per table is limited depending on how

many tables have been created (*see* also **Note 79** and the "CTC-RUO Fixed" application in the user manual).

79. At maximum 10 routing groups can be created. The total number of cells to be isolated is limited depending on the number of groups and the total number of cells in the groups (e.g., with one routing group a maximum of 105 cells can be selected; with two routing groups 51 cells per group can be selected, etc.; for more details see user manual). The name of the groups can be changed by right-click if necessary.

80. The "Cell Panel" consists of the "Sidebar" (selection of groups and parameters), the "Action Bar" (information about cell numbers, icons to switch to "Cell Browser" or "Cell Routing"), and the "Panel Grid" (information and pictures of selected cells). It is saved automatically and opens by clicking "Cell Panel" on the top right of the screen in the CellBrowser™ software. Refer to user manual for adaption of the "Cell Panel."

81. "Park Routing" (i.e., moving the cells from their place to the parking chamber of the DEPArray™ cartridge) can be started by using the "Recovery Manager™" interface. This software allows to watch the cells moving (due to the incorporated live camera), to control the exit and recovery of the samples and to assign samples to the "Recovery Support" layout.

82. It is not necessary but recommended to check if air bubbles are at the entrance of the parking/exit chamber, which might block the routing ways. If air bubbles are visible, it is important to contact the Customer Support before starting the park routing, otherwise important cells might be lost.

83. It is possible to route all cells within a group (cell pools) or to select single cells. The selected groups/cells for parking are labeled with a blue dot.

84. The duration of park routing depends on the number of selected cells. Both progress and time left are displayed in the "Park" window.

85. Even if for the "CTC-RUO Fixed" application, the "200 μL-tube-rack" is preset and recommended for maximum purity of the samples, another "Recovery Support" (e.g., if more than 33 tubes are needed for one experiment) or other types can be chosen by clicking the "+" field in the "Exit and Recovery" section. Choose the appropriate "Recovery Support" from the drop-down menu and "Add" it. By clicking "Ok" an additional layout becomes visible in the "Recover Support" section. For changing the trays during recovery, follow the instructions in the user manual.

86. "A1" is the defined position for the priming recovery, which is not to be confused with the cell recoveries. Once selected by right-click and "Set Priming Recover Position" it is marked in orange and also displayed in the "Exit and Recovery" section.
87. For collecting the putative CTCs as cell groups or single cells within the "CTC-RUO Fixed" application, it is recommended only to use positions in columns 3, 6, 9, and 12 of the "200 μL-tube-rack" plate layout. This results in a maximum of 33 allowable positions (and tubes), "Priming Recovery" included. It is suggested to use the positions for the tubes starting from column 3 (A3, B3, C3, etc., then columns 6, 9, and 12, accordingly). The positions selected for the recovery are marked in blue and are also displayed in the "Exit and Recovery" section. All cells which have not been selected for recovery yet, stay in the "Unassigned" group. Refer to user manual for more detailed information.
88. In order to ensure purity of the samples to be recovered, it is recommended to set a "Blank Recovery" between the different cell types by right-clicking the position in the plate layout and "Set Blank Recover." The position is marked in blue and is also displayed in the "Exit and Recovery" section. The "Exit and Recovery" section also indicates how many drops are used for filling each tube.
89. The "Recovery Support" is optimized for the 200 μL tubes of the company *Applied Biosystems*. If other tubes are to be used, it is necessary to get confirmation by the customer support.
90. The lid of the tube at position "A1" should be opened and directed toward "1" (printed on the recovery tray). The same holds true for the tube in position "A3." All other tubes should have their lids directed toward the letters "A"–"H" (printed on the recovery tray). For a detailed instruction, follow the user manual.
91. Once washing is started, it is not possible to switch back to the CellBrowser™ and no more cells can be parked in the parking area.
92. The process can be stopped by any user regardless of who had signed in for starting the machine.
93. DEPArray™ cartridges can be used only once.
94. For saving the data, the connection to the backup unit is obligatory. The backup unit has to be switched on, otherwise the backup will fail. For more details, follow the user manual.
95. The duration of the backup process depends on the amount of data to be saved. If a new run is performed the same day, the backup process can be cancelled by clicking the red button with the cross at the software screen. It is obligatory to switch off

the device and wait for a minimum of 30 s before restarting it for a new run. The created data which are not saved yet in the backup unit are saved in an internal storage place. They will be saved during the next backup run and will not be lost.

96. The volume of the tubes with 21–85 cells differs from the volume of the tubes with 1–20 cells. The drop rules for the cell recovery are defined depending in which recovery format (e.g., 200 μL tube rack) the cells are collected (for more details see user manual).
97. If 1–20 cells are recovered in one tube, centrifuge for 30 s at 14,100 × *g*. If cell pools with 21–85 cells are recovered in one tube, centrifuge for 10 min at 14,100 × *g*. Make sure that all tube caps are aligned equally with regard to the center of the rotor. It is important to know where exactly the cells will be located after centrifugation in order not to lose the cells when reducing the supernatant to 1 μL in **step 8**.
98. This step is very critical in terms of losing the collected cells. The handling is improved by experience.

References

1. Miller MC, Doyle GV, Terstappen LW (2010) Significance of circulating tumor cells detected by the cellsearch system in patients with metastatic breast colorectal and prostate cancer. J Oncol 2010:617421. https://doi.org/10.1155/2010/617421
2. Alberter B, Klein CA, Polzer B (2016) Single-cell analysis of CTCs with diagnostic precision: opportunities and challenges for personalized medicine. Expert Rev Mol Diagn 16(1):25–38. https://doi.org/10.1586/14737159.2016.1121099
3. Polzer B, Medoro G, Pasch S, Fontana F, Zorzino L, Pestka A, Andergassen U, Meier-Stiegen F, Czyz ZT, Alberter B, Treitschke S, Schamberger T, Sergio M, Bregola G, Doffini A, Gianni S, Calanca A, Signorini G, Bolognesi C, Hartmann A, Fasching PA, Sandri MT, Rack B, Fehm T, Giorgini G, Manaresi N, Klein CA (2014) Molecular profiling of single circulating tumor cells with diagnostic intention. EMBO Mol Med 6(11):1371–1386. https://doi.org/10.15252/emmm.201404033
4. Cristofanilli M, Budd GT, Ellis MJ, Stopeck A, Matera J, Miller MC, Reuben JM, Doyle GV, Allard WJ, Terstappen LW, Hayes DF (2004) Circulating tumor cells, disease progression, and survival in metastatic breast cancer. N Engl J Med 351(8):781–791. https://doi.org/10.1056/NEJMoa040766
5. de Bono JS, Scher HI, Montgomery RB, Parker C, Miller MC, Tissing H, Doyle GV, Terstappen LW, Pienta KJ, Raghavan D (2008) Circulating tumor cells predict survival benefit from treatment in metastatic castration-resistant prostate cancer. Clin Cancer Res 14(19):6302–6309. https://doi.org/10.1158/1078-0432.CCR-08-0872
6. Cohen SJ, Punt CJ, Iannotti N, Saidman BH, Sabbath KD, Gabrail NY, Picus J, Morse M, Mitchell E, Miller MC, Doyle GV, Tissing H, Terstappen LW, Meropol NJ (2008) Relationship of circulating tumor cells to tumor response, progression-free survival, and overall survival in patients with metastatic colorectal cancer. J Clin Oncol 26(19):3213–3221. https://doi.org/10.1200/JCO.2007.15.8923
7. Bidard FC, Peeters DJ, Fehm T, Nole F, Gisbert-Criado R, Mavroudis D, Grisanti S, Generali D, Garcia-Saenz JA, Stebbing J, Caldas C, Gazzaniga P, Manso L, Zamarchi R, de Lascoiti AF, De Mattos-Arruda L, Ignatiadis M, Lebofsky R, van Laere SJ, Meier-Stiegen F, Sandri MT, Vidal-Martinez J, Politaki E, Consoli F, Bottini A, Diaz-Rubio E, Krell J, Dawson SJ, Raimondi C, Rutten A, Janni W, Munzone E, Caranana V, Agelaki S, Almici C, Dirix L, Solomayer EF, Zorzino L, Johannes H, Reis-Filho JS, Pantel K, Pierga JY, Michiels S (2014) Clinical validity of circulating

tumour cells in patients with metastatic breast cancer: a pooled analysis of individual patient data. Lancet Oncol 15(4):406–414. https://doi.org/10.1016/S1470-2045(14)70069-5

8. Rack B, Schindlbeck C, Juckstock J, Andergassen U, Hepp P, Zwingers T, Friedl TW, Lorenz R, Tesch H, Fasching PA, Fehm T, Schneeweiss A, Lichtenegger W, Beckmann MW, Friese K, Pantel K, Janni W, Group SS (2014) Circulating tumor cells predict survival in early average-to-high risk breast cancer patients. J Natl Cancer Inst 106(5). https://doi.org/10.1093/jnci/dju066

Chapter 3

Isolation of Viable Epithelial and Mesenchymal Circulating Tumor Cells from Breast Cancer Patients

Justyna Topa, Anna J. Żaczek, and Aleksandra Markiewicz

Abstract

Circulating tumor cells (CTCs) undergoing epithelial-mesenchymal transition (EMT) may exhibit more aggressive features than epithelial CTCs and are more frequently observed during disease progression. Therefore, detection and characterization of both epithelial and mesenchymal CTCs in cancer patients are urgently needed to allow for a better understanding of the metastatic process and more effective treatment. Here we describe a method for detection and isolation of viable epithelial and mesenchymal CTCs from peripheral blood of breast cancer patients. The method is based on density gradient centrifugation, multiplex immunofluorescent staining, and negative anti-CD45 selection. Cells obtained after the procedure are suitable for genomic or transcriptomic profiling, and they can also be isolated by micromanipulation for single-cell analysis.

Key words Breast cancer, Circulating tumor cells, Epithelial-mesenchymal transition, Mesenchymal phenotype

1 Introduction

Breast cancer (BC) is the most common cancer type in women worldwide and there is an upward trend in the number of BC cases in recent years [1]. A prognosis for BC patients strictly depends on the disease advancement at the time of diagnosis [2] and molecular subtype [3]. Distant metastases remain the main cause of BC-related death, and metastasis mediators, circulating tumor cells (CTCs) present in the bloodstream, are a factor of poor prognosis, in both early and metastatic BC patients [2, 4–6].

CTCs are very heterogeneous, as they may change their phenotype epithelial-mesenchymal transition (EMT) [7]. In this process, BC cells lose their epithelial characteristics (apico-basal polarity, strong adhesion to other cells and basal membrane) and gain aggressive features (motility, ability to degrade extracellular matrix, features of stem cells) [8–10]. It has been shown that mesenchymal CTCs are associated with shorter overall survival

Miodrag Gužvić (ed.), *Single Cell Analysis: Methods and Protocols*, Methods in Molecular Biology, vol. 2752,
https://doi.org/10.1007/978-1-0716-3621-3_3,

and unfavorable tumor characteristics [6, 7], and are more frequently present in the advanced stages of the disease [11]. During disease progression, cancer cells may change their phenotype into the most favorable to particular conditions due to EMT and its reverse process—mesenchymal-epithelial transition (MET) [12, 13]; activation of EMT might promote the dissemination, whereas reversing the mesenchymal phenotype might be crucial for establishing new metastases [14].

Many methods allowing CTCs enrichment, detection, and isolation were already described [15–17], with a vast majority relying on epithelial markers (including golden standard in CTCs detection—CellSearch®), limiting their ability to detect cells undergoing EMT, and further characterization of mesenchymal CTCs. The method described here involves immunofluorescent (IF) staining of peripheral blood mononuclear cells (PBMCs) fraction isolated from BC patient's blood by density gradient centrifugation and negative selection of CD45-positive blood cells, which is an unbiased way of enrichment of CTCs with epithelial and mesenchymal features [6, 7]. For IF staining, we used antibodies directed against common epithelial cell-surface markers, EpCAM and E-cadherin; as well as MCAM (CD146), a protein present on the surface of EpCAM-negative BC cell lines [18]. MCAM is associated with mesenchymal phenotype and increased motility of BC cell lines [19] and has been shown to improve CTCs detection in BC patients [20]. As MCAM may be present on circulating endothelial cells [21], the method includes CD31 (endothelial marker) in the panel of exclusion markers, next to CD45 (blood cells marker) and DAPI (dead cells marker). Our approach (*see* Subheading 2) allows observation of epithelial markers in the "green" channel, whereas mesenchymal markers in the "red" channel. All exclusion markers (CD31, CD45, DAPI) are visible in the "blue" channel. This setup may be freely modified by the addition of other identified CTCs markers. Performing IF staining before blood cells depletion minimizes accidental loss of rare cancer cells during multiple washing steps. CD45-positive cells negative selection by Dynabeads™ CD45 is one of the most efficient methods for CTCs enrichment regardless of their phenotype [17, 22] and in combination with IF staining allows high recovery of living cells (Fig. 1a). The described method also enables high enrichment of CTCs fraction, as after the whole procedure about 98.4% of blood cells are depleted (Fig. 1b) when 5 mL of blood was processed.

2 Materials

2.1 CTCs Enrichment

1. Low bind 1.5 mL tubes (e.g., Ultra High Recovery Tubes, Starlab, cat. No. E1415-2600 or Protein LoBind Tubes, Eppendorf, cat. No. 0030108116), 15 mL and 50 mL tubes.

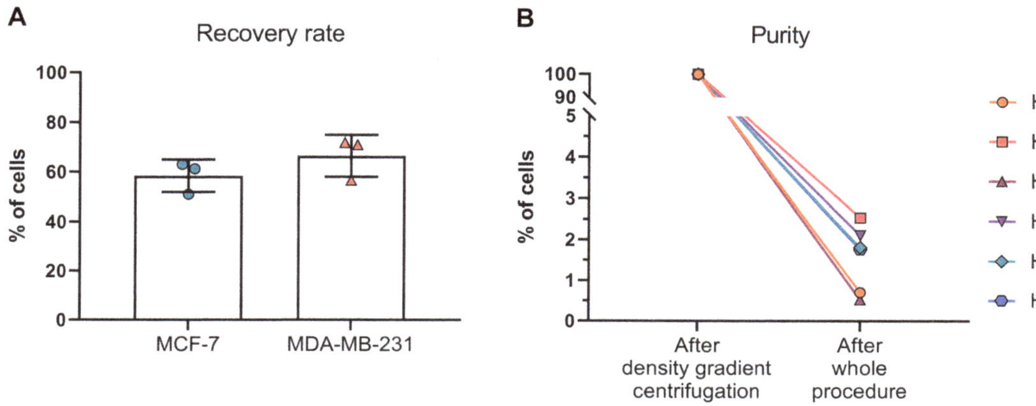

Fig. 1 (**a**) The recovery rate of 100 MCF-7 (epithelial) cells and 100 MDA-MB-231 (mesenchymal) cells of BC cell lines spiked into 5 mL of blood. Recovery rate is shown as mean ± SD from three independent experiments. (**b**) Number of nucleated cells after density gradient centrifugation and after the whole procedure as determined by processing 5 mL of blood from healthy donors (with no diagnosed cancer) samples ($n = 6$)

2. K_2EDTA-coated blood collection tubes (BD Vacutainer®).
3. Syringes and syringe filters with 0.22 μm pores.
4. Coating Buffer: 2 mM EDTA, 1% FBS in 1xPBS (pH 7.4). Measure out 1 L of 1xPBS and transfer into 1 L glass bottle. Weigh 0.744 g of disodium EDTA 2-hydrate and transfer into a bottle. Mix on a roller or magnetic stirrer until EDTA dissolves. Sterilize in autoclave. After cooling down, add 10 mL of sterile FBS, mix gently, and divide into 100 mL portions (*see* **Note 1**). Store at 4 °C for up to 4 weeks.
5. Histopaque®-1077 (Sigma Aldrich).

2.2 Immunofluorescent Staining

1. Blocking Buffer: 50 mM glycine, 5% BSA in 1xPBS (pH 7.4). Measure out 100 mL of 1xPBS and transfer into 100 mL glass bottle. Weigh 0.375 g of glycine and 5 g of BSA. Transfer into a bottle and mix on a roller or magnetic stirrer until glycine and BSA dissolve. Perform filtration through filters with 0.22 μm pores. Divide into 1 mL portions, leaving 10 mL for Staining Buffer preparation. Store at −20 °C for up to 6 months.
2. Staining Buffer: 10 mM glycine, 1% BSA in 1xPBS (pH 7.4). Measure out 10 mL of sterile Blocking Buffer and transfer into 50 mL bottle. Refill to the volume of 50 mL with 1xPBS. Divide into 1 mL portions and store at −20 °C for up to 6 months.
3. Anti-EpCAM antibody: clone VU1D9, Alexa Fluor® 488-conjugated, Cell Signalling Technology, cat. No. 5198 (*see* **Note 2**).

4. Anti-E-cadherin antibody: clone 67A4, Alexa Fluor® 488-conjugated, Santa Cruz Biotechnology, cat. No. sc-21791.
5. Anti-MCAM antibody—clone P1H12, Alexa Fluor® 594-conjugated, Santa Cruz Biotechnology, cat. No. sc-18837.
6. Anti-CD31 antibody—clone WM-59, Super Bright 436-conjugated, eBioscience, cat. No. 62-0319-42.
7. Anti-CD45 antibody—clone REA747, VioBlue®-conjugated, Miltenyi Biotec, cat. No. 130-110-637.
8. DAPI—stock 1 mg/mL, intermediate stock 1:200 (stored at −20 °C, aliquoted).
9. Sample Mixer (e.g., HulaMixer® Sample mixer, Thermofisher).

2.3 CD45-Positive Cells Negative Selection

1. Dynabeads™ CD45, Thermofisher, cat. No. 11153D.
2. DynaMag™-15 Magnet, Thermofisher, cat. No. 12301D.

2.4 Additional Equipment

1. Centrifuge with temperature, acceleration, and brake control, and with tilting rotor, for 15 and 50 mL tubes.
2. Centrifuge with temperature control for 1.5 mL tubes.

3 Methods

The CTCs isolation method is based on density gradient centrifugation, multiplex IF staining, and anti-CD45 depletion (Fig. 2).

3.1 CTCs Enrichment

1. Prepare all buffers and reagents needed during the procedure: Coating Buffer, 1xPBS, Histopaque® at room temperature, and additional Coating Buffer and 1× PBS in 4 °C (*see* **Note 3**). Pre-coat 2× low-bind 1.5 mL tubes, 4 × 15 mL, and 1 × 50 mL tubes; add 1, 2, and 10 mL of Coating Buffer (at room temperature) into tubes, respectively, and put on a roller for 15 min to allow coating walls of the tubes with the buffer and prevent cells sticking to the tube. After 15 min remove the solution from the tubes.
2. By venipuncture collect blood sample into K_2EDTA-coated tubes. Discard the tube with the first 1 mL of blood that might contain contaminating epithelial cells or fibroblasts due to skin punctuation. Use a second K_2EDTA-coated tube to collect 5 mL of blood for the analysis (*see* **Note 4**).
3. To remove platelets, transfer blood into a pre-coated 15 mL tube and centrifuge at 200 × *g*, 21 °C for 10 min. Discard the

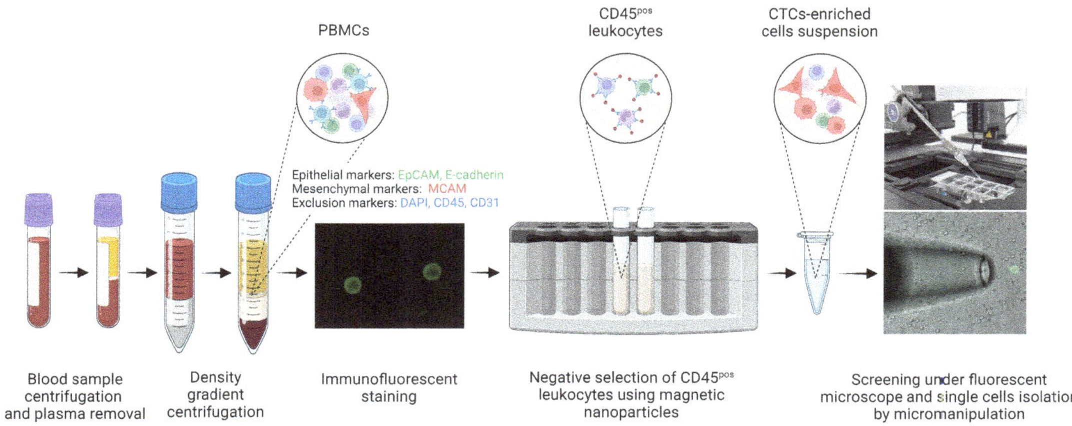

Fig. 2 The workflow of the CTCs isolation method

top layer of plasma and avoid taking any of the cellular fraction beneath it (around ⅔ can easily be removed).

4. Refill the remaining sample with 1xPBS (room temperature) to the volume of 9 mL, gently mix, and carefully layer onto 4 mL of sterile Histopaque® in the new pre-coated 15 mL tube. Be careful not to mix the layers of diluted blood and Histopaque®, as it may affect the separation of cells. Centrifuge the samples at 400 × *g*, 21 °C for 30 min, with no brake and acceleration (*see* **Note 5**).
5. *From that moment carry out all the steps on ice.* Collect PBMCs fraction (*see* **Note 6**) into new pre-coated 15 mL tube and refill to the volume of 10 mL with cold (4 °C) 1xPBS. Centrifuge at 450 × *g*, 4 °C, 10 min to obtain a cell pellet. Discard the supernatant and immediately proceed to the next steps of the procedure.

3.2 Immuno-fluorescent Staining and Negative Selection of CD45-Positive Cells

1. Suspend the pellet in 250 μl of Blocking Buffer (prepared according to the description in the Subheading 2.2) and transfer it into a pre-coated 1.5 mL tube. To wash out any remaining cells (possibly also CTCs), rinse the 15 mL tube that contained PBMCs fraction with another 250 μl of Blocking Buffer and transfer it into the 1.5 mL tube containing suspended cell pellet. Incubate at 4 °C for 15 min on the sample mixer. Centrifuge at 400 × *g*, 4 °C, 5 min.
2. Carefully remove the supernatant with a pipette (*see* **Note 7**) and add 200 μl of freshly prepared antibodies mix (Table 1) into the pellet, with gentle pipetting. *From that moment sample must be protected from light.* Incubate sample at 4 °C for 30 min on sample mixer.

Table 1
Antibodies mix used for immunofluorescent staining[a]

Component	Final dilution	Volume
anti-EpCAM Ab	1:200	1
anti-E-cadherin Ab	1:100	2.0
anti-MCAM Ab	1:200	1.0
anti-CD31 Ab	1:100	2.0
DAPI	1:10,000	2.0
Staining buffer	–	192

[a]Does not include CD45 antibody, which is added later during CD45-positive cells depletion

3. During staining of the PBMCs fraction, prepare Dynabeads™ CD45 for further use. Transfer 125 μl of well-resuspended Dynabeads™ CD45 into a pre-coated 15 mL tube (*see* **Note 8**). Add 1 mL of Coating Buffer and mix with Pasteur pipette. Place the tube into the DynaMag™-15 Magnet for 1 min and then discard the supernatant. Be careful not to touch magnetic beads on the side of the tube. Take the tube off the magnet and resuspend washed Dynabeads™ CD45 in 125 μl of Coating Buffer.
4. When IF staining of PBMCs fraction is finished, centrifuge the tube at 400 × *g*, 4 °C, 1 min, then gently suspend the cells, and transfer into the pre-coated 15 mL tube containing washed Dynabeads™ CD45. To collect residual cells from the tube, rinse the 1.5 mL tube after PBMCs IF staining with 675 μl of Coating Buffer and transfer it to a 15 mL tube with PBMCs and Dynabeads™ CD45 (total volume should be 1 mL). Incubate the sample for 5 min at 4 °C on the sample mixer. During the incubation, CD45-positive cells are captured with anti-CD45 antibody-coated magnetic nanoparticles (*see* **Note 9**).
5. Add anti-CD45 antibody directly to the tube (10 μl, to the final dilution of 1:100) with PBMCs and Dynabeads™ CD45. After 25 min of incubation (at 4 °C), remove the tube from the mixer and dilute the sample to the volume of 13 mL with cold (4 °C) Coating Buffer (*see* **Note 10**). Mix sample with Pasteur pipette and place tube in the DynaMag™-15 Magnet for 10 min (cover the magnet with the tube to protect it from light) CD45-positive cells attached to the magnetic beads remain on the side of the tube while kept on the magnet, whereas CD45-negative cells remain in the solution.
6. Carefully transfer CTCs-enriched cells suspension into a pre-coated 50 mL tube and fill it to the volume of 25 mL

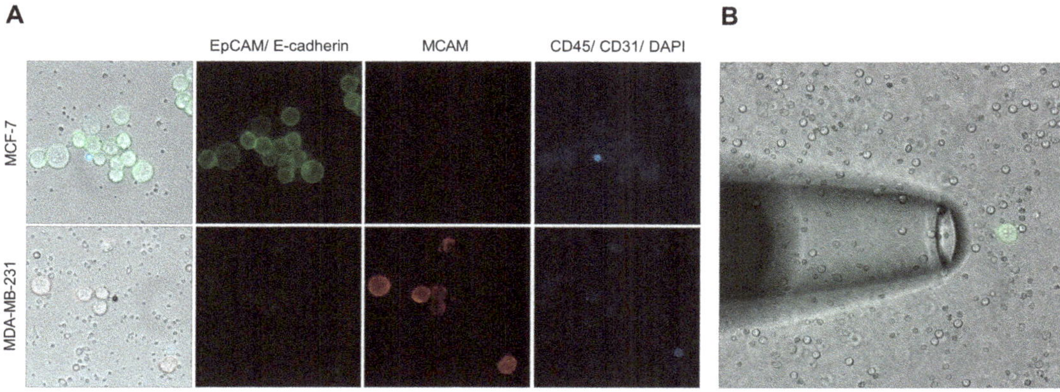

Fig. 3 (**a**) Spiked-in MCF-7 cells positive for epithelial markers (EpCAM and E-cadherin; green), and MDA-MB-231 positive for MCAM (red) after CTCs isolation procedure. Spiked cells show high viability (DAPI-negative) and are negative for CD45 and CD31 (blue). (**b**) Picking of single MCF-7 cell positive for epithelial markers

with cold (4 °C) Coating Buffer. Centrifuge at 400 × *g*, 4 °C for 5 min and remove supernatant with a pipette controller or Pasteur pipette (*see* **Note 11**), leaving about 100 μl of the liquid above the pellet. Resuspend the pellet and transfer it to a pre-coated 1.5 mL tube. To collect any remaining cells, rinse the bottom of the 50 mL tube with 100–200 μl of Coating Buffer and transfer it into 1.5 mL tube containing CD45-depleted CTCs fraction.

3.3 Single CTCs Isolation by Micromanipulation

Such prepared CTCs-enriched cell suspension may be further processed as bulk or used for single-cell analyses. For our research, single CTCs were isolated by micromanipulation (Fig. 3) described elsewhere in this volume. Such captured single cells may be further subjected to genomic and transcriptomic analyses, as demonstrated in other protocols in this volume.

4 Notes

1. Add FBS under the laminar hood, to ensure the buffer remains sterile. It will allow longer storage of the Coating Buffer. 100 mL of the buffer is enough to carry out the CTCs isolation procedure from two 5 mL blood samples.
2. Fluorochromes should be matched to the lasers and filters available on a particular microscope.
3. About 25 mL of 1xPBS (10 mL at RT and 15 mL at 4 °C) and 50 mL of Coating Buffer (20 mL at RT and 30 mL at 4 °C) are enough to perform the procedure on one 5 mL blood sample. Histopaque® may be aliquoted into 4 mL portions, to reach room temperature faster.

4. Collected blood should be processed as soon as possible, preferably within 2 h after collection. After blood collection, the sample should be stored at 4–8 °C, but it should be left to equilibrate to room temperature for density gradient centrifugation.
5. Ensure that acceleration and brake are turned off to avoid mixing the layers after density gradient centrifugation.
6. Crucial step: ensure that the whole PBMCs fraction is collected. It is better to aspirate a higher volume of Histopaque® or diluted plasma than leave uncollected cells, among which may be CTCs.
7. You can leave about 50 μl of the supernatant above the cell pellet, to avoid accidental cells aspiration.
8. It is crucial to mix well Dynabeads™ CD45, to aspirate the proper volume of Ab-coated magnetic particles. In this procedure, we are using an excess of the reagent to ensure efficient CD45-positive cells depletion from 5 mL of blood.
9. During immunomagnetic depletion, IF staining with anti-CD45 antibody is performed. However, to avoid blocking the CD45 molecule from binding immunomagnetic anti-CD45 beads, first magnetic beads are added, followed by a fluorescently labeled anti-CD45 antibody.
10. Make sure to aspirate also the solution that might be present in the cap of the tube.
11. Supernatant may be aspirated with the use of glass Pasteur pipette connected to vacuum pump, but attention should be paid to not aspirate cell pellet.

Acknowledgments

This work was supported by the National Science Centre grant number 2016/21/D/NZ3/02629.

References

1. Sung H, Ferlay J, Siegel RL, Laversanne M, Soerjomataram I, Jemal A, Bray F (2021) Global cancer statistics 2020: GLOBOCAN estimates of incidence and mortality worldwide for 36 cancers in 185 countries. CA Cancer J Clin 71(3):209–249. https://doi.org/10.3322/caac.21660
2. Dong G, Wang D, Liang X, Gao H, Wang L, Yu X, Liu J (2014) Factors related to survival rates for breast cancer patients. Int J Clin Exp Med 7(10):3719–3724
3. Fallahpour S, Navaneelan T, De P, Borgo A (2017) Breast cancer survival by molecular subtype: a population-based analysis of cancer registry data. CMAJ Open 5(3):E734–E739. https://doi.org/10.9778/cmajo.20170030
4. Bidard FC, Peeters DJ, Fehm T, Nole F, Gisbert-Criado R, Mavroudis D, Grisanti S, Generali D, Garcia-Saenz JA, Stebbing J, Caldas C, Gazzaniga P, Manso L, Zamarchi R, de Lascoiti AF, De Mattos-Arruda L, Ignatiadis M, Lebofsky R, van Laere SJ, Meier-Stiegen F, Sandri MT, Vidal-

Martinez J, Politaki E, Consoli F, Bottini A, Diaz-Rubio E, Krell J, Dawson SJ, Raimondi C, Rutten A, Janni W, Munzone E, Caranana V, Agelaki S, Almici C, Dirix L, Solomayer EF, Zorzino L, Johannes H, Reis-Filho JS, Pantel K, Pierga JY, Michiels S (2014) Clinical validity of circulating tumour cells in patients with metastatic breast cancer: a pooled analysis of individual patient data. Lancet Oncol 15(4):406–414. https://doi.org/10.1016/S1470-2045(14)70069-5

5. Cristofanilli M, Budd GT, Ellis MJ, Stopeck A, Matera J, Miller MC, Reuben JM, Doyle GV, Allard WJ, Terstappen LW, Hayes DF (2004) Circulating tumor cells, disease progression, and survival in metastatic breast cancer. N Engl J Med 351(8):781–791. https://doi.org/10.1056/NEJMoa040766
6. Markiewicz A, Nagel A, Szade J, Majewska H, Skokowski J, Seroczynska B, Stokowy T, Welnicka-Jaskiewicz M, Zaczek AJ (2018) Aggressive phenotype of cells disseminated via hematogenous and lymphatic route in breast cancer patients. Transl Oncol 11(3):722–731. https://doi.org/10.1016/j.tranon.2018.03.006
7. Markiewicz A, Topa J, Nagel A, Skokowski J, Seroczynska B, Stokowy T, Welnicka-Jaskiewicz M, Zaczek AJ (2019) Spectrum of epithelial-mesenchymal transition phenotypes in circulating tumour cells from early breast cancer patients. Cancers (Basel) 11(1). https://doi.org/10.3390/cancers11010059
8. Kalluri R, Weinberg RA (2009) The basics of epithelial-mesenchymal transition. J Clin Invest 119(6):1420–1428. https://doi.org/10.1172/JCI39104
9. Mani SA, Guo W, Liao MJ, Eaton EN, Ayyanan A, Zhou AY, Brooks M, Reinhard F, Zhang CC, Shipitsin M, Campbell LL, Polyak K, Brisken C, Yang J, Weinberg RA (2008) The epithelial-mesenchymal transition generates cells with properties of stem cells. Cell 133(4):704–715. https://doi.org/10.1016/j.cell.2008.03.027
10. Moreno-Bueno G, Portillo F, Cano A (2008) Transcriptional regulation of cell polarity in EMT and cancer. Oncogene 27(55):6958–6969. https://doi.org/10.1038/onc.2008.346
11. Zhang S, Wu T, Peng X, Liu J, Liu F, Wu S, Liu S, Dong Y, Xie S, Ma S (2017) Mesenchymal phenotype of circulating tumor cells is associated with distant metastasis in breast cancer patients. Cancer Manag Res 9:691–700. https://doi.org/10.2147/CMAR.S149801
12. Gao X, Liu X, Lu Y, Wang Y, Cao W, Liu X, Hu H, Wang H (2019) PIM1 is responsible for IL-6-induced breast cancer cell EMT and stemness via c-myc activation. Breast Cancer 26(5):663–671. https://doi.org/10.1007/s12282-019-00966-3
13. Ocana OH, Corcoles R, Fabra A, Moreno-Bueno G, Acloque H, Vega S, Barrallo-Gimeno A, Cano A, Nieto MA (2012) Metastatic colonization requires the repression of the epithelial-mesenchymal transition inducer Prrx1. Cancer Cell 22(6):709–724. https://doi.org/10.1016/j.ccr.2012.10.012
14. Markiewicz A, Topa J, Popeda M, Szade J, Skokowski J, Welnicka-Jaskiewicz M, Zaczek A (2021) Activation of epithelial-mesenchymal transition process during breast cancer progression—the impact of molecular subtype and stromal composition. Acta Biochim Pol 68(3):385–392. https://doi.org/10.18388/abp.2020_5719
15. Bailey PC, Martin SS (2019) Insights on CTC biology and clinical impact emerging from advances in capture technology. Cell 8(6). https://doi.org/10.3390/cells8060553
16. Rushton AJ, Nteliopoulos G, Shaw JA, Coombes RC (2021) A review of circulating tumour cell enrichment technologies. Cancers (Basel) 13(5). https://doi.org/10.3390/cancers13050970
17. Topa J, Grešner P, Żaczek AJ, Markiewicz A (2022) Breast cancer circulating tumor cells with mesenchymal features—an unreachable target? Cell Mol Life Sci 79(2):81. https://doi.org/10.1007/s00018-021-04064-6
18. Mostert B, Kraan J, Bolt-de Vries J, van der Spoel P, Sieuwerts AM, Schutte M, Timmermans AM, Foekens R, Martens JW, Gratama JW, Foekens JA, Sleijfer S (2011) Detection of circulating tumor cells in breast cancer may improve through enrichment with anti-CD146. Breast Cancer Res Treat 127(1):33–41. https://doi.org/10.1007/s10549-010-0879-y
19. Zabouo G, Imbert AM, Jacquemier J, Finetti P, Moreau T, Esterni B, Birnbaum D, Bertucci F, Chabannon C (2009) CD146 expression is associated with a poor prognosis in human breast tumors and with enhanced motility in breast cancer cell lines. Breast Cancer Res 11(1):R1. https://doi.org/10.1186/bcr2215
20. Onstenk W, Kraan J, Mostert B, Timmermans MM, Charehbili A, Smit VT, Kroep JR, Nortier JW, van de Ven S, Heijns JB, Kessels LW, van Laarhoven HW, Bos MM, van de Velde CJ, Gratama JW, Sieuwerts AM, Martens JW, Foekens JA, Sleijfer S (2015) Improved circulating tumor cell detection by a combined EpCAM and MCAM cellsearch enrichment approach in

patients with breast cancer undergoing neoadjuvant chemotherapy. Mol Cancer Ther 14(3): 821–827. https://doi.org/10.1158/1535-7163.MCT-14-0653

21. Strijbos MH, Gratama JW, Kraan J, Lamers CH, den Bakker MA, Sleijfer S (2008) Circulating endothelial cells in oncology: pitfalls and promises. Br J Cancer 98(11):1731–1735. https://doi.org/10.1038/sj.bjc.6604383

22. Guo M, Li X, Zhang S, Song H, Zhang W, Shang X, Zheng Y, Jiang H, Lv Q, Jiang Y, Hao H (2015) Real-time quantitative RT-PCR detection of circulating tumor cells from breast cancer patients. Int J Oncol 46(1):281–289. https://doi.org/10.3892/ijo.2014.2732

Chapter 4

Single-Cell Recovery from Tumor Cell Xenotransplanted Zebrafish Embryos for the Study of Metastasis-Initiating Cells

Pablo Hurtado, Inés Martínez-Pena, and Roberto Piñeiro

Abstract

The study of metastasis-competent cells at the single-cell level represents an opportunity to decipher the molecular mechanisms associated with the metastatic cascade as well as to understand the functional and molecular heterogeneity of these cells. In this context, preclinical in vivo models of cancer metastasis are valuable tools to understand the behavior of cancer cells throughout the process. Here we describe a detailed protocol for the isolation and recovery of individual viable human metastatic cells from zebrafish embryos xenotransplanted with cancer cells for downstream molecular analysis. We cover the critical steps for the dissociation of the xenografted zebrafish embryos to generate a single-cell suspension, and the micromanipulation for their recovery as single cells.

Key words Metastasis, Zebrafish, Dissociation, Single-cell suspension, Micromanipulation

1 Introduction

The metastatic cascade is a complex multistep process by which cancer cells abandon the primary tumor and reach distant organs and tissues where they grow into new tumors. Cancer metastasis is a highly inefficient process, and only a very small proportion of the cells released by the primary tumor are able to survive this process and successfully seed metastases [1]. The potential of cancer cells to form metastases is mainly determined by their genetic composition as well as by the interactions with the tumor microenvironment (TME).

The use of animal models, and in particular the mouse, has been an important tool to dissect the different stages of the metastatic process and unravelling their interaction with the TME along the metastatic process [2]. In recent years, the zebrafish embryo has emerged as an alternative for murine xenograft models of metastasis, allowing a more reliable dissection of the metastatic cascade and

Miodrag Guzvic (ed.), *Single Cell Analysis: Methods and Protocols*, Methods in Molecular Biology, vol. 2752,
https://doi.org/10.1007/978-1-0716-3621-3_4,

the study of the metastatic competency of cancer cells [3–5]. Their fast extra-uterine development, their transparency, and the lack of a mature adaptive immune system make them a very good organism to model tumor growth and metastasis [3]. Therefore, zebrafish xenografts allow to study in real time the behavior of just a few hundred cancer cells and to follow them on time individually.

The recent development of methods and technologies to study cancer cells at a single-cell level has opened new avenues to, among others, characterize the intra-tumor cellular heterogeneity, and to identify rare cancer cell subpopulations [6], e.g., metastasis-initiating cells. Thus, nowadays it is possible to capture and isolate individual cancer cells from different sources, such as tumor tissue or blood biopsies, to study their mutation and gene expression profile. The combination of these technologies with in vivo preclinical models of cancer metastasis will make plausible to decipher the molecular mechanisms underlying the metastatic competence of tumor-disseminated cells under controlled conditions.

Here we describe a protocol for the isolation of individual metastatic tumor cells from xenotransplanted zebrafish embryos for the development of downstream molecular assays (e.g., DNA or RNA analyses). This protocol explains in detail the different steps for the efficient dissociation of the zebrafish embryo in order to generate a single-cell suspension. It describes how to manipulate and dissect the zebrafish embryos to obtain the tissue sections where cells have disseminated, and the following dissociation process. The dissociation process combines an enzymatic and mechanical digestion that allows to obtain a homogeneous single-cell suspension. In addition, the protocol includes a detailed step-by-step description of the micromanipulation process to successfully obtain a pure and viable cell for molecular interrogation. It describes the main considerations to take into account (e.g., cell density and purity) for an easy and efficient recovery of the cells of interest without carrying over contaminating material present in the sample such as host cells. It also describes a few control tips to ensure the adequate integrity of the collected cell as well as the efficient transfer to the final recipient tube. Importantly, this protocol can be carried out using a manual micromanipulator and a mechanical microinjector, without the need to use expensive automated equipment. Lastly, this protocol is suitable for the isolation of intact viable cells, preserving nucleic acid integrity and allowing to obtain genetic material compatible with downstream molecular assays such as RNA gene expression analysis.

2 Materials

2.1 Reagents

2.1.1 Zebrafish Embryo Euthanasia and Dissociation

1. Sterile PBS 1× (w/o Ca^{++} and Mg^{++}).
2. 0.4% Tricaine (ethyl 3-aminobenzoate methanesulfonate salt) solution in water.
3. 0.25% Trypsin-EDTA solution.
4. 100 mg/mL Collagenase Type I from *Clostridium histolyticum* solution in PBS.
5. Fetal Bovine Serum (FBS).
6. Cell culture medium (e.g., DMEM, RPMI, etc.).

2.1.2 Single-Cell Micromanipulation

1. Sterile PBS 1× (w/o Ca^{++} and Mg^{++}).
2. 5% Bovine serum albumin (BSA) w/v solution in PBS.

2.2 Equipment

2.2.1 Zebrafish Embryo Euthanasia and Dissociation

1. Surgical blade.
2. Plastic Pasteur pipette (3 mL).
3. Micropipettes 2–20 μl, 20–200 μl, and 200–1000 μl.
4. Stereo microscope.
5. Heat block.

2.2.2 Single-Cell Micromanipulation

1. Reaction tube (1.5 mL).
2. 0.2 mL thin-walled PCR tube, sterile.
3. Micropipette 0.2–2 μl.
4. Chambered cell culture slide, sterile.
5. Centrifuge capable of spinning down 0.2 mL PCR tubes.
6. Borosilicate thin wall capillaries, 1.0 mM OD, 0.78 mM ID, 75 mM.
7. Manual micromanipulator.
8. Mechanical microinjector.
9. Inverted fluorescence microscope.

3 Methods

The following protocol is designed to analyze individual cancer cells (disseminated or metastatic cells) isolated from the tails of five-day post-fertilization (dpf) embryos, although it could potentially be applied for any cell found in the embryo's body. Xenotransplanted tumor cells should be previously stained with fluorescent lipid markers or transgenically engineered to express fluorescent proteins (e.g., GFP and mCherry) in order to follow their behavior inside the fish body and allow an easy discrimination between them and the host cells throughout the process.

3.1 Zebrafish Embryo Euthanasia and Dissection (see Note 1)

1. Prepare a 0.4% tricaine solution (anesthetic): dissolve 400 mg of tricaine in 97.9 mL of ddH_2O. Add 2.1 mL of Tris–HCl 1 M pH 9 solution. Adjust to pH 7.0. Keep the solution at 4 °C for short-term use and at −20 °C for long-term storage.
2. Dilute the 0.4% tricaine solution ten times in water to a final concentration of 0.04% (tricaine overdose) and add 2 mL of the solution to a small beaker.
3. Prepare a reaction tube (1.5 mL) adding 1 mL of PBS w/o Ca^{+} $^{+}$ and Mg^{++}.
4. Remove the xenotransplanted fish (≈15 zebrafish embryos at a developmental stage of 5 days post fertilization) from the incubation water using a 3 mL plastic Pasteur pipette (*see* **Note 2**) and place them in the beaker containing the tricaine solution to induce the dead by overdose of anesthetic.
5. Incubate the fish in this solution for 1–2 min, until they are completely anesthetized (*see* **Note 3**).
6. Transfer one embryo at a time with the help of a plastic Pasteur pipette to a clean dissection surface (e.g., microscopy glass slide) and place it under a stereo microscope to proceed with the dissection of the embryo's body (*see* **Note 4**). Make sure to add a few drops of the anesthetic solution to the fish to avoid the sample from drying out (*see* **Note 5**).
7. With the help of fine-tip tweezers and a scalpel or razor blade, orientate the fish as desired and physically separate the fish yolk sac from the tail, where the cancer cells are found disseminated (Fig. 1, **step 1**, and box) (*see* **Note 6**).
8. Cut the tail into smaller pieces in order to improve the digestion efficiency. Collect all tail pieces from the dissection surface and transfer it to the PBS-containing reaction tube for washing.
9. Repeat this process with the rest of fish and sequentially transfer the dissected tails to the reaction tube.
10. Invert the tube a few times to ensure that the tissues are washed (Fig. 1, **step 2**).
11. With the help of a micropipette (200–1000 μl), carefully remove the PBS from the tube avoiding the collection of the embryo's pieces (Fig. 1, **step 3**).
12. Repeat the washing step once more.

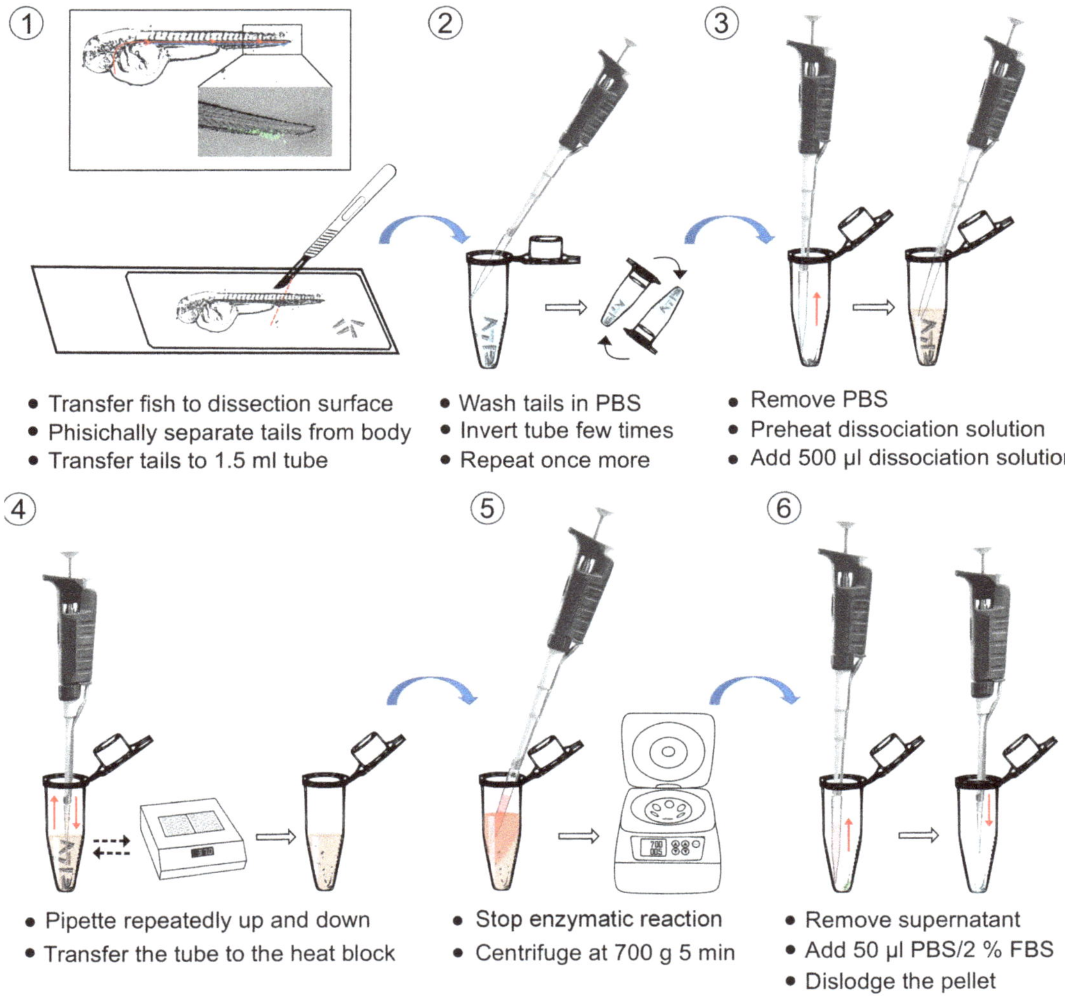

Fig. 1 Schematic representation of the protocol for zebrafish embryo dissociation. 1. Zebrafish embryo dissection. 2. Washing of dissected tissue. 3–4. Enzymatic and mechanical digestion of the tissues. 5. Stopping the enzymatic reaction and pelleting of the cells. 6. Obtaining the single cell suspension

3.2 Zebrafish Embryo Enzymatic and Mechanical Digestion

1. Dissolve the collagenase in PBS at a final concentration of 100 mg/mL. This solution can be stored at −20 °C in small aliquots for later use.
2. Prepare 500 µl of the dissociation solution by mixing 20 µl of collagenase (stock 100 mg/mL) and 480 µl of 0.25% Trypsin-EDTA solution in a tube and keep it at 37 °C in a heat block.
3. Add the 500 µl of dissociation solution previously heated to the reaction tube containing the tissues and transfer it to a heat block at 37 °C (Fig. 1, **step 3**) (*see* **Note** 7).
4. With the help of a 200–1000 µl micropipette, every 2 min pipette repeatedly up and down for a few times and vortex

the sample to dislodge the cells. Transfer the tube to the heat block in between mechanical processing (Fig. 1, **step 4**) (*see* **Note** 8).

5. Incubate the sample for no longer than 10 min at 37 °C.
6. Prepare 5 mL of cell culture medium with 10% FBS by mixing in a sterile tube 4.5 mL of medium and 500 μl of FBS.
7. Once the incubation time is finished, add 800 μl of cell culture medium with 10% of FBS with a micropipette to neutralize the action of the dissociation solution and avoid cell damage (Fig. 1, **step 5**).
8. Transfer the tube to a bench-top centrifuge and spin down the cells by centrifuging at 700 g for 5 min at room temperature (Fig. 1, **step 5**).
9. Carefully remove the supernatant and add 50 μl of PBS + 2% FBS to the cellular pellet and dislodge it with a micropipette (Fig. 1, **step 6**). Keep the cell suspension at room temperature and avoid putting them in ice.
10. Optional step. A DNA stain for live cells (e.g., Hoechst 33342) can be added to the cell suspension to visualize the nuclear morphology and integrity. For use, follow the manufacturer's instructions.
11. Visualize cells under the microscope to check for the presence of fluorescent cells (tumor cells) and their morphology.

3.3 Preparation of Collection Chamber for a Cellular Suspension

1. Prepare a 5% BSA w/v solution. Weigh 500 mg of BSA and dissolve it in 10 mL of 1× PBS. Mix well until the BSA is fully dissolved. Filter the solution with a 0.22 μm pore. Add 100 μl of the 5% BSA solution at every well of a sterile chambered cell culture slide and remove it quickly (*see* **Note** 9).
2. Differentiate within the slide at least two chambers, one as the picking field and the other one (or others) as the sample field (Fig. 2, **step 1**).
3. Add 200 μl of PBS in each picking field (*see* **Note** 10).
4. Add 190 μl of PBS in the sample field plus 10 μl of the cell suspension (Fig. 2, **step 1**). It is important that the maximum concentration of cells in the chamber is not superior to 1×10^3 cells/chamber. It will ensure an efficient single-cell isolation without contaminating cells being carried over (Fig. 3) (*see* **Note** 11).
5. Place the chambered cell culture slide in the fluorescence microscope specimen holder.
6. Allow cells to set at the bottom of the sample field for a few minutes.

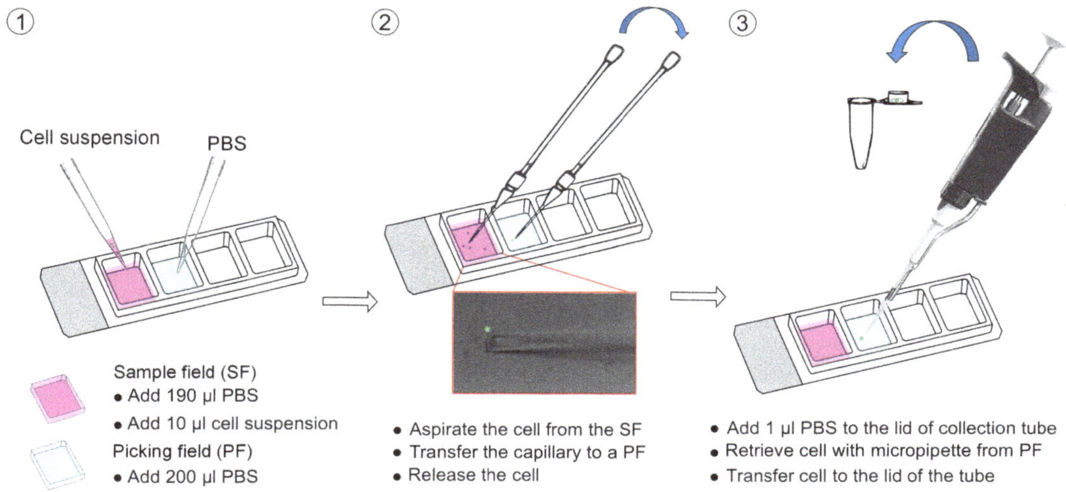

Fig. 2 Scheme depicting the main steps of the process of micromanipulation

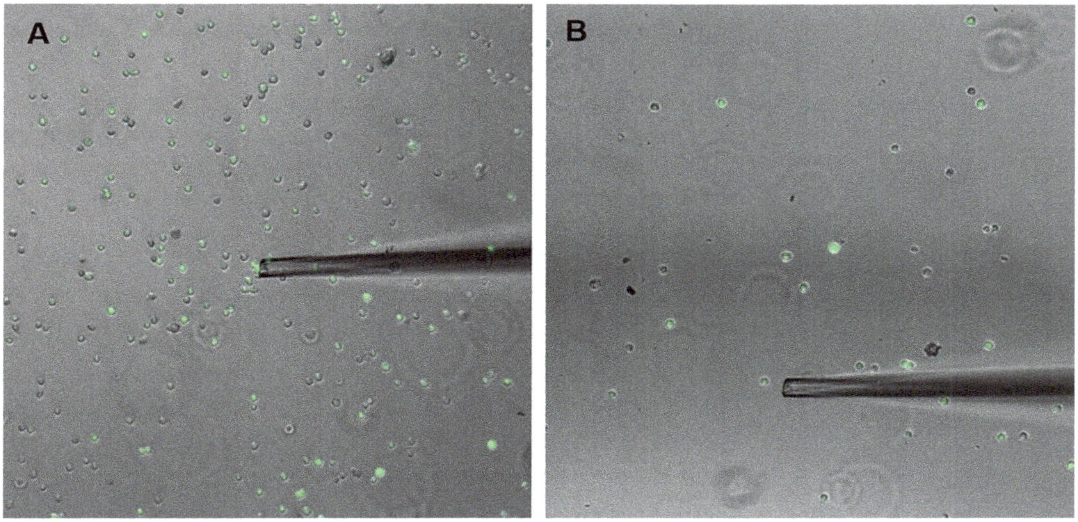

Fig. 3 Requirement of cellular density at the sample field for single-cell micromanipulation. Representative images of a too-high cellular density (**a**) and an adequate cellular density (**b**) at sample field

7. Check under the microscope that all cells are sitting at the bottom of the slide (*see* **Note** 12).

3.4 Capillary Preparation for Micromanipulation

1. Add 100 µl of FBS to a sterile 1.5 mL reaction tube.
2. Immerse the capillary inside the reaction tube and let the FBS to enter the tip by capillarity.
3. Insert the capillary into the capillary holder of the microinjector, and place it onto the mechanical arm of the micromanipulator (Fig. 4) (*see* **Note** 13).

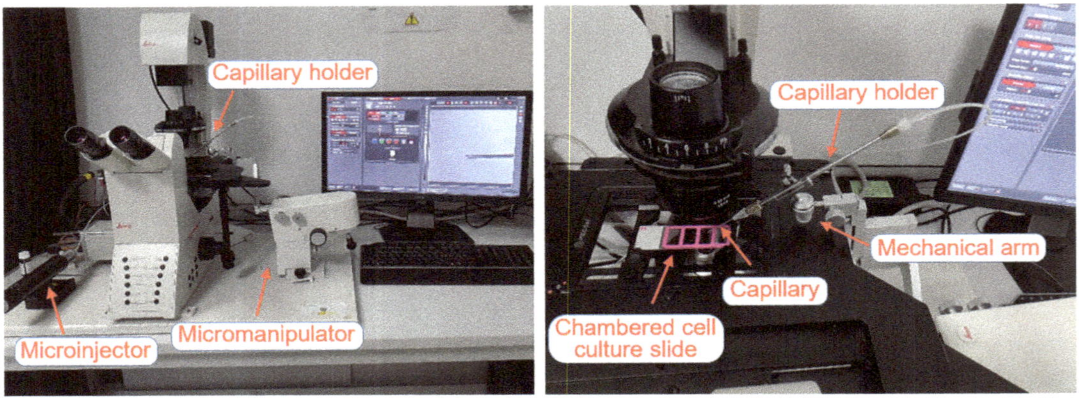

Fig. 4 Equipment setup for single-cell micromanipulation. This protocol is described for the use of a mechanical microinjector and a manual micromanipulator, as shown in the images

4. With the micromanipulator control knobs, immerse the capillary into the sample field and bring it to the bottom, softly touching the surface of the slide (*see* **Note** 14). For easy visualization, use the 10× objective of the microscope; this will allow an efficient focus of the capillary.
5. With the capillary inside the sample field, expulse a small volume of FBS from the capillary and absorb few microliters of PBS, making use of the mechanical microinjector control knob. Always make sure that the capillary is never void of any liquid inside.

3.5 Cell Picking

Before beginning with the micromanipulation step, make sure all the necessary setup for the micromanipulation is previously done (Fig. 4).

1. Add 1 μl of PBS to the lid of a 0.2 mL sterile PCR tube (*see* **Note** 15). This tube will be used to collect the retrieved cell from the cell suspension.
2. Use a 20× microscope objective to focus the capillary and for the aspiration of the cells from the sample field.
3. Locate the cell of interest within the sample field (*see* **Note** 16) and carefully approach the capillary to it making use of the control knobs. Place the tip of the capillary opposite to the cell of interest taking the precaution of not pushing or flushing it away.
4. Clean the area surrounding the cell of interest by flushing away the contaminating cells with the capillary using the microinjector control knob. This can be easily done by aspirating and flushing PBS with the capillary, which will wash away the possible contamination.

5. Proceed to aspirate the cell into the capillary. The disappearance of the cell from the vision field should be verified before retrieving the capillary.
6. Move the capillary away from the sample field, transfer it into the picking field, and release the cell (Fig. 2, **step 2**) (*see* **Note** 17).
7. Check under the microscope the presence and purity of the cell. In many occasions, the cell of interest is still accompanied by contaminating material carried over from the sample field (i.e., non-wanted zebrafish cells or debris that could interfere with the downstream molecular analysis).
8. In the event that the cell is surrounded by contaminating material, clean the surroundings as indicated in point **4** to ensure that the cell remains alone (*see* **Note** 18).
9. Completely remove the capillary from the picking field.
10. Visually check for the integrity of the cell and the purity of the area surrounding it. Take a second photograph of the cell.
11. Slowly rest the tip of a 0.2–2 μl micropipette on the surface of the slide, next to the cell. Carefully retrieve the cell with the micropipette previously set up to aspirate 1 μl (Fig. 2, **step 3**) (*see* **Note** 19).
12. Transfer the 1 μl containing the cell to the lid of the 0.2 mL PCR tube previously prepared by immersing the tip in the liquid and releasing the cell inside (Fig. 2, **step 3**) (*see* **Note** 20).
13. Place the tube's lid on the fluorescence microscope and check for the presence of the cell (*see* **Note** 21).
14. Once the presence of the cell is confirmed, close the tube lid and spin it down for a short time to bring the liquid to the bottom of the tube.
15. Store the tube at −80 °C to preserve the integrity of the genetic material or alternatively store it under the desired conditions for downstream analysis.

4 Notes

1. It is convenient prior to the euthanasia and dissection of the xenotransplanted embryos to verify under a fluorescence microscope the presence of the desired tumor cell population in the fish tails as well as to locate them within the tails. This would help to minimize the size of the dissected tissue for a more efficient digestion.

2. Alternatively, a 200 μl micropipette tip can be cut off to collect the embryos. Cut the tip wide enough to avoid damaging the fish when manipulating them.
3. The addition of fish to the anesthetic solution should be done sequentially. Avoid exposing the fish long times to the anesthesia to preserve the viability of the cells.
4. The stereo microscope allows for a higher magnification and better view and definition of the area of the fish to dissect.
5. Dry zebrafish tissue becomes very sticky and difficult to remove from the glass slide surface.
6. This protocol is designed for the isolation of tumor cells disseminated from the yolk sack or duct of Cuvier (injection sites) to the tail of 5 days post-fertilization (dpf) embryos (metastatic cells).
7. Alternatively, a water bath or incubator set at 37 °C can be used for the incubation.
8. This procedure allows for a better enzymatic and mechanical disassociation of the embryo's tissues.
9. BSA helps to avoid cell attachment to the surface of the slides.
10. The volume in the chamber can be modified for comfort, but it is convenient to have enough volume inside the chamber.
11. The total number of cells will depend on the size of the digested tissue, the efficiency of the digestion, and the volume in which the cell pellet is resuspended.
12. Make sure at this point that the dilution of the cells is adequate for cell isolation. Dilute the sample if needed.
13. This protocol is described for the use of a manual micromanipulator and a mechanical microinjector (Fig. 4), although automated versions could be used. Follow the manufacturer's instructions for the setup of the equipment.
14. As glass capillaries are flexible, they do not break when softly touch the surface of the slide.
15. Instead of PBS, cell lysis buffer or another desired storage buffer can be used at this point.
16. It is advisable to take a photograph of the cell before picking it to contrast it with a photograph taken later in the process of manipulation. It will allow to assess any possible damage exerted on the cell or change on morphology.
17. Release only a small volume of PBS into the picking field to avoid flushing the retrieved cell away from the field of view.
18. On occasions, the cell of interest can be attached to other cells due to a lack of efficiency during tissue digestion. The capillary

can be used as a blade to cut off possible unions with other cells.

19. A micropipette is used to retrieve the cell instead of the capillary to minimize the risk of contamination, as undesired cells or other material could be trapped inside the capillary. An alternative to the use of the micropipette would be the use of a new clean capillary. However, removing the old capillary and attaching a new one for every cell to be collected will substantially increase the duration of the procedure, compromising the quality of the biological material.
20. Immersing the tip in the liquid guarantees the efficient transfer of the cell to the collection tube.
21. Placing the liquid in the lid of the tube, rather than at the bottom, facilitates the visualization of the cell under the microscope.

Acknowledgments

This work was supported by Roche-Chus Joint Unit (IN853B 2018/03), funded by Axencia Galega de Innovación (GAIN), Consellería de Economía, Emprego e Industria. I.M.-P. is funded by the Training Program for Academic Staff fellowship (FPU16/01018), from the Ministry of Education and Vocational Training, Spanish Government.

References

1. Bystricky B, Mego M (2016) Circulating tumor cells in breast cancer patients. Neoplasma 63(1): 18–29. https://doi.org/10.4149/neo_2016_003
2. Gomez-Cuadrado L, Tracey N, Ma R, Qian B, Brunton VG (2017) Mouse models of metastasis: progress and prospects. Dis Model Mech 10(9):1061–1074. https://doi.org/10.1242/dmm.030403
3. Hill D, Chen L, Snaar-Jagalska E, Chaudhry B (2018) Embryonic zebrafish xenograft assay of human cancer metastasis. F1000Res 7:1682. https://doi.org/10.12688/f1000research.16659.2
4. Mercatali L, La Manna F, Groenewoud A, Casadei R, Recine F, Miserocchi G, Pieri F, Liverani C, Bongiovanni A, Spadazzi C, de Vita A, van der Pluijm G, Giorgini A, Biagini R, Amadori D, Ibrahim T, Snaar-Jagalska E (2016) Development of a patient-derived xenograft (PDX) of breast cancer bone metastasis in a Zebrafish model. Int J Mol Sci 17(8). https://doi.org/10.3390/ijms17081375
5. Teng Y, Xie X, Walker S, White DT, Mumm JS, Cowell JK (2013) Evaluating human cancer cell metastasis in zebrafish. BMC Cancer 13:453. https://doi.org/10.1186/1471-2407-13-453
6. Saadatpour A, Lai S, Guo G, Yuan GC (2015) Single-cell analysis in cancer genomics. Trends Genet 31(10):576–586. https://doi.org/10.1016/j.tig.2015.07.003

Chapter 5

Isolation of Single Circulating Tumor Cells Using VyCAP Puncher System

Thais Pereira-Veiga, Bianca Behrens, Joska J. Broekmaat, Lisa Oomens, Michiel Stevens, Arjan G. J. Tibbe, Nikolas Stoecklein, Laura Muinelo-Romay, Roberto Piñeiro, and Clotilde Costa

Abstract

Tumor heterogeneity has a major role in the development of tumor evasion and resistance to treatments. To study and understand the intrinsic heterogeneity of cancer cells, the use of single-cell isolation technology has had a major boost in recent years, gaining ground to bulk analysis in the study of solid tumors. In the liquid biopsy field, the use of technologies for single-cell analysis has represented a major advance in the study of the heterogeneity of circulating tumor cells (CTCs), providing relevant information about therapy-resistant CTCs. However, single-cell analysis of CTCs is still challenging due to the weakness and scarcity of these cells. In this chapter, we describe a protocol for CTCs isolation at a single-cell level using the VyCAP Puncher system.

Key words CTC, Liquid biopsy, VyCAP, Puncher, Single cell, Sequencing, CellSearch, Tumor heterogeneity

1 Introduction

Circulating tumor cells (CTCs) have been shown of high clinical relevance, being directly involved in the metastatic process [1]. Although they represent a unique opportunity to study cancer metastasis and progression, they are at a low rate compared with other blood cells. In recent years, a range of technologies for the enrichment of CTCs has emerged to overcome this issue. CellSearch® is the only FDA-approved method for CTCs quantification while Parsortix® has been cleared for use in metastatic breast cancer patients for the capture and harvest of CTCs from whole blood. This enumeration had proved to have prognostic value in metastatic breast, prostate, and colorectal cancer patients [2]. Still, CTC numbers offer limited information and do not allow for

Miodrag Gužvić (ed.), *Single Cell Analysis: Methods and Protocols*, Methods in Molecular Biology, vol. 2752, https://doi.org/10.1007/978-1-0716-3621-3_5,

physician intervention; therefore, the current trend is to analyze genetic data from CTCs [3].

The strategies for CTCs isolation have limitations since specific markers for CTCs have not yet been identified; there is no a clear consensus about the definition of a CTC and the sensitivity of some technologies is still low for addressing their molecular characterization. It should be noted that currently, no system is directly capable of isolating a population of pure CTCs. Thus, the available technologies isolate some normal blood cells from the hematopoietic fraction in a non-specific way together with the CTCs. Therefore, a combination of different isolation strategies is necessary, including a first step of CTC enrichment and subsequent phenotypic characterization of the isolated cells to determine their tumor origin. Single-cell analysis is the best choice for an in-depth analysis of heterogeneity studies as well as to avoid biased results due to unspecific cells. Different technologies are currently in the market for CTC isolation at a single-cell level. DEPArray [4, 5], CELLCELECTOR™ [6], or VyCAP Puncher [7] are the most widely used. Here, we describe the VyCAP Puncher protocol for CTCs isolation and subsequent sequencing analysis. Here, we describe the VyCAP Puncher protocol for CTCs isolation and subsequent sequencing analysis. The starting specimens are CTC pre-enriched samples in CellSearch® MAGNEST® cartridges, from the blood of metastatic cancer patients. The CellSearch® system uses whole blood and captures CTCs with magnetic beads, based on the positive expression of EpCAM. Subsequently, it performs immunofluorescence staining of the enriched fraction of cells to identify those that are positive for cytokeratins and negative for CD45, in combination with round/oval morphology and nucleus/cytoplasm localization. The current CellSearch® system defines a CTC as a positive event when it has a well-defined nucleus (positive DAPI staining), expresses cytokeratin (CK8, CK18, and CK19), does not express CD45, and is larger than 4×4 μm in size.

VyCAP technology separates single cells into single microwells after which the cells can be imaged and isolated on a Puncher system. The system automatically creates images of all cells and allows enumeration of CTCs. Besides, it gives the option of isolating individual cells from a CTC pre-enriched fraction that is compatible with other current technologies (CellSearch® system, Parsortix, RosetteSep, etc.) (Fig. 1). The immunofluorescence staining performed previously in the cells allows the recognition of CTCs in the VyCAP Puncher software. For the isolation of individual cells, the system is composed of two essential parts: a disposable with microwell chip that is used to distribute individual cells in a microwells and a Puncher system that automatically selects and isolates cells. There is no limitation regarding the number of cells that the user can isolate.

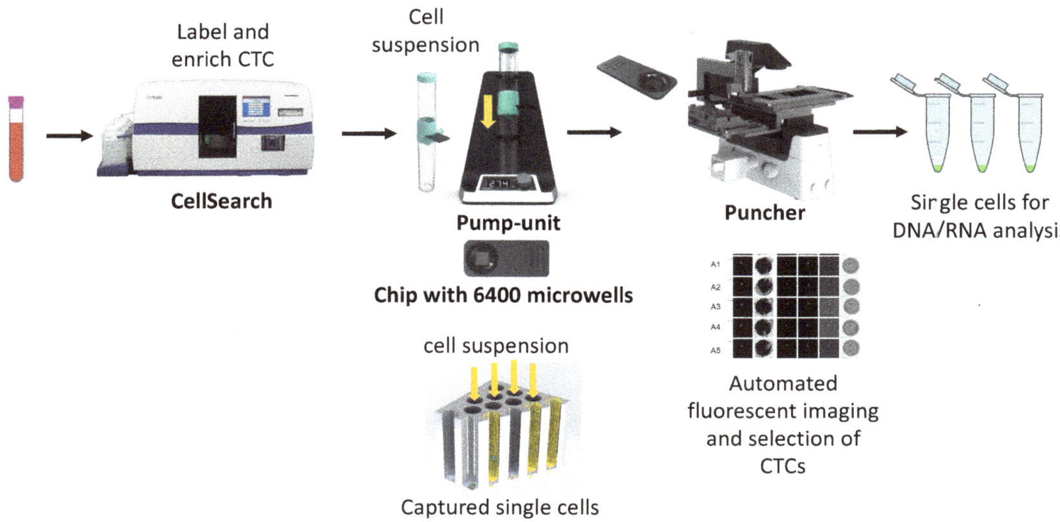

Fig. 1 Workflow scheme for isolation of single CTCs using VyCAP Puncher from a pre-enriched sample by Cellsearch technology

2 Materials

PBS 1× filtered (0.45 μm).

Plastic tubes (1.5 mL).

Long glass Pasteur pipettes or gel loading pipette tips.

Filtered ethanol (0.45 μm).

Mineral oil, embryo culture tested.

Sterile 0.2 mL thin-walled tube, classic semi-domed cap, regular profile.

Centrifuge capable of spinning down 0.2 mL PCR tubes.

3 Methods

3.1 Recovering MAGNEST® Cartridge Content

1. Remove the cartridge from the MAGNEST® cartridge holder after scanning (*see* **Note 1**).
2. Remove the cartridge plug by pushing it upwards from the back of the cartridge (*see* **Note 2**).
3. Homogenize the sample inside the cartridge by pipetting it up and down inside the cartridge at least 10 times, using a Pasteur pipette or a gel-loading pipette tip. This procedure will detach the cells from the cartridge.
4. Transfer the content of the cartridge to a 1.5 mL tube.
5. Using a pipette, add 300 μl of filtered PBS 1× inside the cartridge by leaning the tip against the front side of the emptied cartridge (glass side where the barcode is located). Pipette this

fluid up and down for ten times, rinsing the surface. The removal of the cells can be checked by placing the cartridge glass slide up under a standard upright fluorescence microscope.

6. Add the 300 μl of PBS wash into the same 1.5 mL tube that the sample was previously transferred to. The final sample volume in the 1.5 mL tube is 600 μl.

3.2 Cell Seeding

1. Pre-wet the VyCAP microwell chip by adding 1 mL of filtered 100% ethanol on top of it. Let the chip sit with the ethanol for 45 min.
2. Connect a disposable filtration unit to the pump and insert the microwell chip in the slot.
3. Add 1 mL of filtered PBS 1× to the remaining ethanol that is on top of microwell chip.
4. Set the pressure of the pump station to 10 mbar (*see* **Note 3**).
5. Switch on the pump and pull most of the liquid through. Switch off the pump action when there is only little amount of fluid left. It is utterly important not to remove all fluid as this will lead to air entering the microwells, which will limit the number of available microwells.
6. Rinse the microwell chip by adding another 1 mL of filtered PBS 1× to the chip and pulling most of the liquid through. Repeating this step several times will completely remove the ethanol.
7. Once the microwell chip is rinsed, add 1 mL of filtered PBS 1× to the system and switch on the pump unit.
8. As soon as a drop of PBS has passed the microwells, add the 600 μl of the sample recovered from the MAGNEST® cartridge (*see* **Note 4**).
9. Switch off the pump after the whole volume of the sample has passed.
10. Remove the microwell chip from the filtration station.
11. Wash the back side of the microwell chip with filtered PBS 1×.
12. Dry the back side of the microwell chip gently with a soft tissue without directly touching the surface of the chip.

3.3 Isolation of CTCs with the Punching System

1. Transfer the microwell chip into the Puncher system to acquire fluorescence images.
2. The Puncher system contains the VyCAP Imaging system, an automated image acquisition software. This system has auto focus and an automatic filter cube changer that can hold a maximum of six different filter cubes.

3. After image acquisition, the software presents a pre-selection of the cells of interest, based on the fluorescent cell signature.
4. Pipette 35 μl of mineral oil in the cap of the 0.2 mL tube and place it in the sample holder on the puncher device.
5. Bend the cap so that its top is at the same height as the opening of tube.
6. Select the tube collection format and align the needle to punch into the cap of the tube.
7. Select the desired cell to be punched into the cap of the 0.2 mL tube (*see* **Note 5**).
8. Close the tube upside down, with the sample liquid still at the cap, and vortex the tube upside down for 5 s.
9. Immediately spin down at 10,000 g for 1 min.
10. Add 1 μl of sterile 1× PBS to every sample. Spin down at 10,000 g for 1 min (*see* **Note 6**).
11. Sample can now be stored at 4 °C for up to 1 week prior to quality control and genome wide amplification.

4 Notes

1. Ideally, proceed with this step immediately after scanning with CellSearch®. Alternatively, the CellSearch® cartridges can be stored in darkness, in horizontal position and at 4 °C to be used up to a year after.
2. It is difficult to reach behind the plastic holder to be able to push the plug out of the cartridge. Use a tool, such as a pair of tweezers to push the plug out.
3. Depending on the sample, the pressure can be set higher, but a higher pressure will induce more stress on the cells.
4. The flow rate for PBS without cells is 1–2 mL/min at 10 mbar. As soon as more pores will become occupied by cells, the flow rate will drop. In case all wells are filled, the flow rate will approach zero.
5. After punching, an image of the punched microwell is acquired to ensure that the microwell bottom and cell are removed.
6. **Steps 8–10** must be performed within 10 min after **step 7**.
7. The number of microwells is 6400. Maximum number of cells that you can add is depending on the size of the cells. The diameter of the exit pore of the microwells is 5 microns. Cells that are flexible or that have a small diameter will pass through the pore. If the sample only contains large rigged cells that are unable to pass through the 5 μm pore, the maximum number of cells in the sample volume is 6400.

Acknowledgement

This work was supported by Roche-Chus Joint Unit (IN853B 2018/03), funded by Axencia Galega de Innovación (GAIN), Consellería de Economía, Emprego e Industria.

References

1. Massague J, Obenauf AC (2016) Metastatic colonization by circulating tumour cells. Nature 529(7586):298–306. https://doi.org/10.1038/nature17038
2. Cristofanilli M, Budd GT, Ellis MJ, Stopeck A, Matera J, Miller MC, Reuben JM, Doyle GV, Allard WJ, Terstappen LW, Hayes DF (2004) Circulating tumor cells, disease progression, and survival in metastatic breast cancer. N Engl J Med 351(8):781–791. https://doi.org/10.1056/NEJMoa040766
3. Keller L, Pantel K (2019) Unravelling tumour heterogeneity by single-cell profiling of circulating tumour cells. Nat Rev Cancer 19(10):553–567. https://doi.org/10.1038/s41568-019-0180-2
4. Di Trapani M, Manaresi N, Medoro G (2018) DEPArray system: an automatic image-based sorter for isolation of pure circulating tumor cells. Cytometry A 93(12):1260–1266. https://doi.org/10.1002/cyto.a.23687
5. Reduzzi C, Motta R, Bertolini G, Miodini P, Martinetti A, Sottotetti E, Daidone MG, Cappelletti V (2017) Development of a protocol for single-cell analysis of circulating tumor cells in patients with solid tumors. Adv Exp Med Biol 994:83–103. https://doi.org/10.1007/978-3-319-55947-6_4
6. Neumann MH, Schneck H, Decker Y, Schomer S, Franken A, Endris V, Pfarr N, Weichert W, Niederacher D, Fehm T, Neubauer H (2017) Isolation and characterization of circulating tumor cells using a novel workflow combining the CellSearch ((R)) system and the CellCelector(). Biotechnol Prog 33(1):125–132. https://doi.org/10.1002/btpr.2294
7. Stevens M, Oomens L, Broekmaat J, Weersink J, Abali F, Swennenhuis J, Tibbe A (2018) VyCAP's puncher technology for single cell identification, isolation, and analysis. Cytometry A 93(12):1255–1259. https://doi.org/10.1002/cyto.a.23631

Chapter 6

Simultaneous Isolation and Amplification of mRNA and Genomic DNA of a Single Cell

Miodrag Gužvić

Abstract

Many biological or pathological processes are driven by cells difficult to identify or isolate, i.e., rare cells. Very often, these cells have elusive biology. Therefore, their detailed characterization is of utmost importance. There are many approaches that allow analysis of few or even many targets within one class of biomacromolecules/analytes (e.g., DNA, RNA, proteins, etc.) in single cells. However, due to rarity of the cells of interest, there is a great need to comprehensively analyze multiple analytes within these cells, in other words to perform multi-omics analysis. In this chapter, I describe a method to isolate, separate, and amplify total mRNA and genomic DNA of a single cells, using whole transcriptome (WTA) and whole genome amplification (WGA). These WTA and WGA products enable simultaneous analysis of transcriptome and genome of a single cell using various downstream high-throughput approaches.

Key words Single-cell analysis, Whole genome amplification, Whole transcriptome analysis, Multi-omics analysis, Rare cells

1 Introduction

Analysis of single cells is becoming a routine approach in biomedical laboratory. In some cases, this type of analysis is enabling is a deeper and more detailed insight into the biological phenomena that we are already familiar with (e.g., comprehensive single-cell transcriptome sequencing enabled unprecedented insight into the heterogeneity of bone marrow populations [1]). In other cases, single-cell analysis is a necessity to understand the mechanisms driven by rare cells, whose biology is largely unknown (e.g., biology of disseminated cancer cells that are founder cells of lethal metastasis [2]). Analyzing individual aspects of the biology of these rare cells (e.g., genome or transcriptome) is often not enough to paint a complete molecular profile of these cells. Therefore, there is a need to develop and use multiple-omics approaches to analyze single-cell biology [3]. These approaches enable us to comprehensively and simultaneously profile various biomacromolecules (e.g., RNA,

Miodrag Gužvić (ed.), *Single Cell Analysis: Methods and Protocols*, Methods in Molecular Biology, vol. 2752, https://doi.org/10.1007/978-1-0716-3621-3_6,

DNA, proteins) or their various functional and structural states (e.g., methylome, exome, chromatin states) [3]. In this chapter, I present a method to isolate and amplify total mRNA and genomic DNA of a single cell. The procedure has two major parts, whole transcriptome amplification (WTA; nota bene: the name of the method traditionally contains the term "transcriptome" although it only captures mRNA, and not other expressed RNA species), and whole genome amplification (WGA). Both methods have been developed to profile disseminated cancer cells [4–6] in humans, but have also been used for other types of human [7, 8] or murine cells [9–11]. Generated WGA and WTA products are quite versatile and can be further analyzed by various methods, like expression microarrays [12], array-CGH [2, 13, 14], end-point PCR [2] and qPCR [15, 16], genome and transcriptome sequencing [17], etc. WTA is based on the capture of mRNAs on the solid phase [6, 12], where reverse transcription (RT) takes place, followed by second cDNA strand synthesis and global amplification (Fig. 1a). WGA is based on fragmentation of the genome using a restriction enzyme [4, 18], ligation of adapters, which enable generation of binding sites for primers that drive the final step of global amplification (Fig. 1b).

2 Materials

2.1 Isolation of mRNA and Genomic DNA, and Whole Transcriptome Amplification

Ideally, this procedure should be performed under laminar flow bench in a room where no DNA amplification takes place and no amplified DNA is stored. During execution of the protocol, gloves should be exchanged frequently, especially upon touching anything that is outside of the bench (face, chair, drawer handles, etc.). It is advised not to wear wristwatches, rings, bracelets, etc.

2.1.1 Reagents and Kits

All reagents should be molecular biology grade, and DNase and RNAse free. All dilutions and mixtures should be made with nuclease-free water. Reagent dilutions made from powder should be filtered using 0.22 μm filter. All dilutions should be aliquoted to avoid frequent freeze-thaw cycles.

- mTRAP kit (Active Motif), containing lysis buffer, protease, streptavidin-coated beads, and biotinylated poly-T gripNA Probe (peptide nucleic acid, PNA) oligonucleotides. Beads are stored at 4 °C, the rest is stored at −20 °C. Individual reagents should be prepared according to manufacturer's instructions (lysis buffer and beads are ready to use, protease and PNAs are lyophilized and need to be dissolved).
- Nuclease-free water.
- Linear polyacrylamide carrier (e.g., AM9520 from Ambion, *see* **Note 1**). Stored at 4 °C.

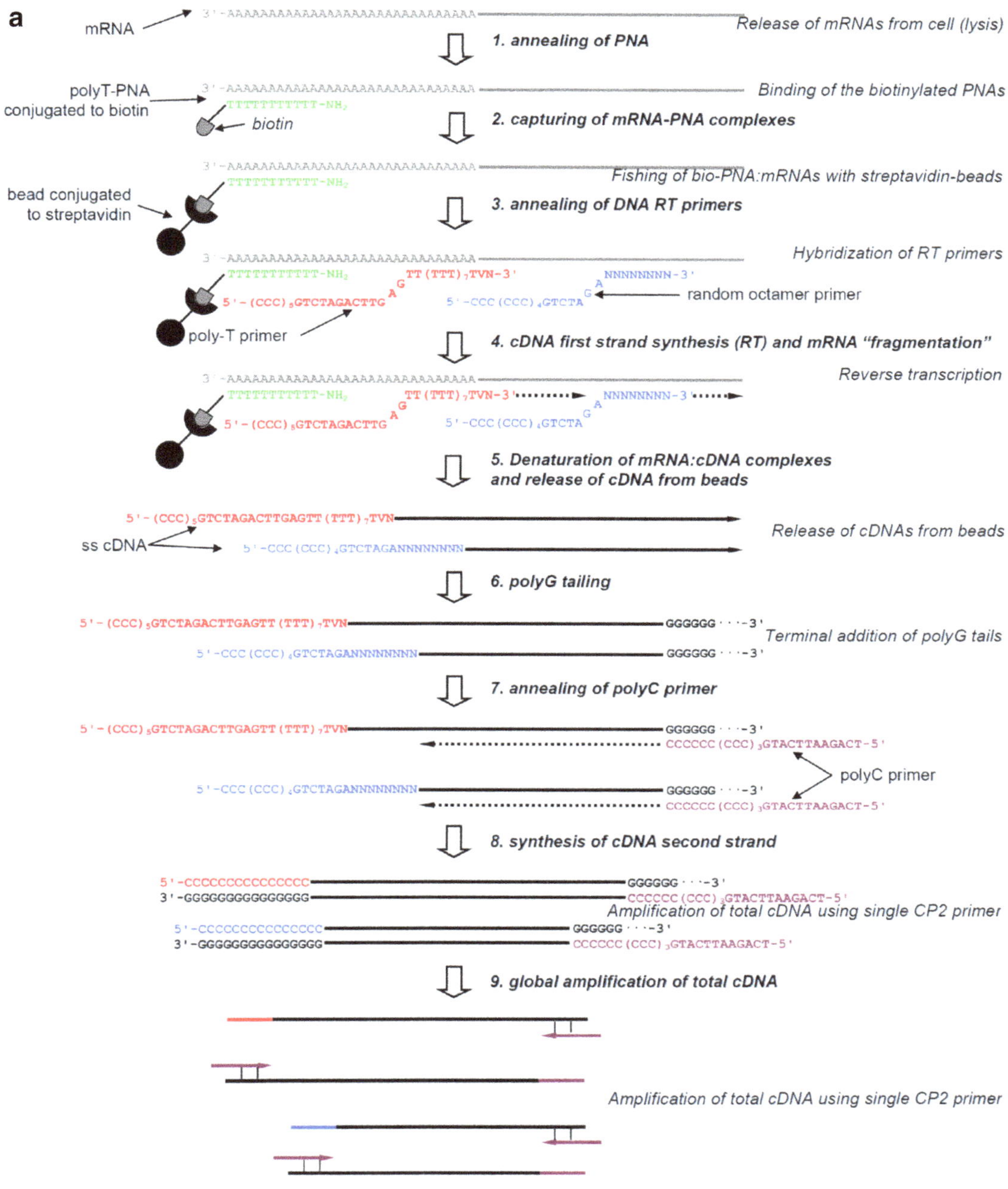

Fig. 1 Schematic representation of WTA procedure (**a**) and WGA procedure (**b**). Please note that the oligonucleotide sequences shown do not have to match those presented in the protocol; the figure serves to display the principles of individual steps

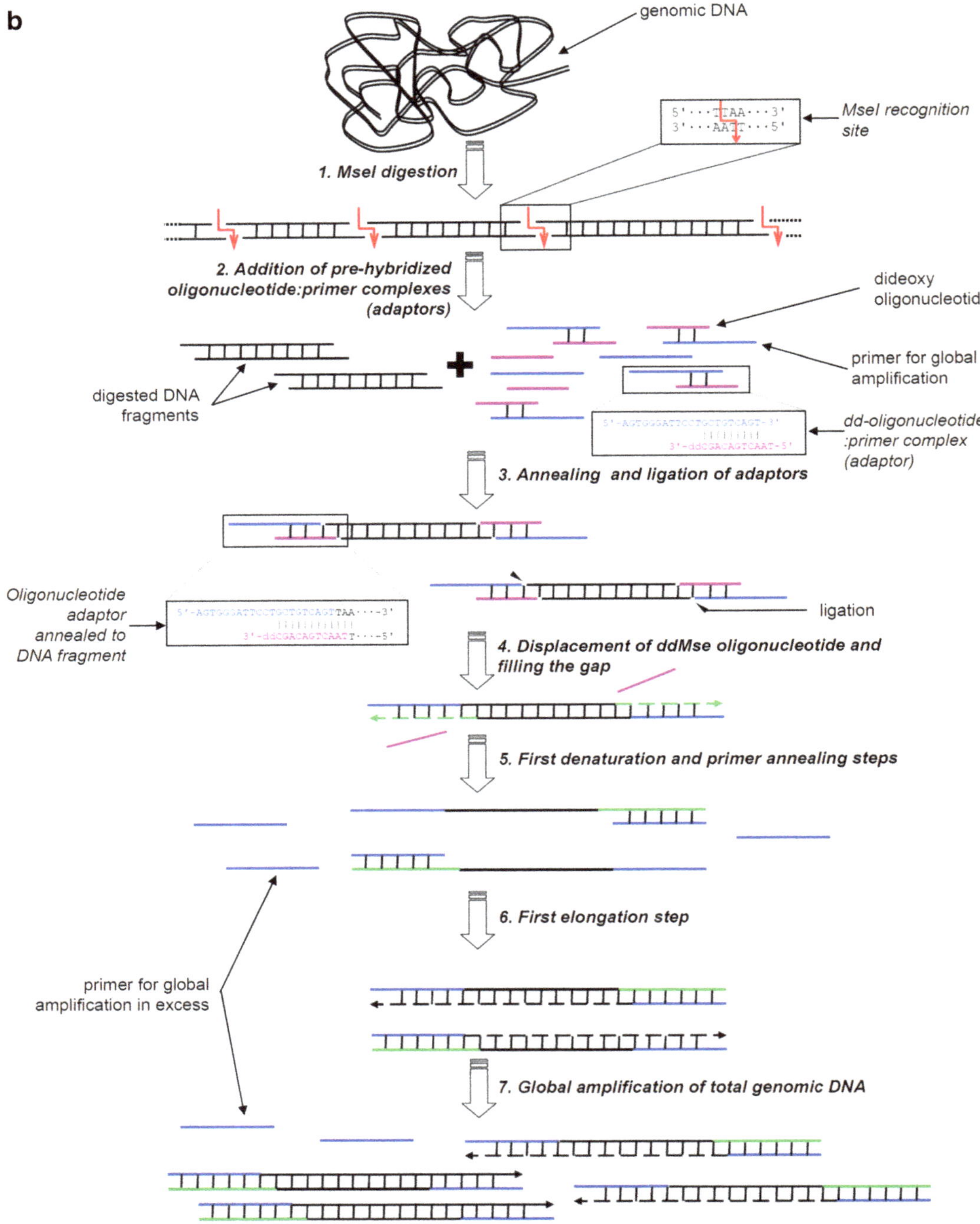

Fig. 1 (continued)

- SuperScript II reverse transcriptase (SSIIRT) kit (Thermo) containing 5× RT buffer, 0.1 M DTT, and SS II reverse transcriptase (RT). Stored at −20 °C (*see* **Note 2**).
- 10% IGEPAL CA-630 in nuclease-free water. Stored at −20 °C.
- dNTP mix in nuclease-free water, 10 mM each. Store at −20 °C.
- Oligonucleotide N8: 5′-CCC CCC CCC CCC CCC GTC TAG ANN NNN NNN-3′ (N = A/C/T/G)—mixture of oligonucleotides containing random octamer sequences. 200 mM dilution. Store at −20 °C (*see* **Note 3**).
- Oligonucleotide T24: 5′- CCC CCC CCC CCC CCC GTC TAG ACT TGA GTT TTT TTT TTT TTT TTT TTT TTT TVN-3′ (V = A/C/G; N = A/C/T/G)—mixture of oligonucleotides containing random 3′ termini. 100 mM dilution. Store at −20 °C (*see* **Note 4**).
- IGEPAL-wash buffer: 50 mM Tris–HCl, pH 8.0, 75 mM KCl, 10 mM DTT, 0.25% IGEPAL CA-630. Store at −20 °C.
- Tween-wash buffer: 50 mM Tris–HCl, pH 8.0, 75 mM KCl, 10 mM DTT, 0.50% Tween 20. Store at −20 °C.
- Ethanol, absolute.
- $MgCl_2$, 40 mM.
- KH_2PO_4, 200 mM. Store at −20 °C.
- DTT, 1 mM (dilute 0.1 M DTT from SSIIRT kit).
- dGTP, 2 mM. Store at −20 °C.
- Terminal deoxynucleotidyl Transferase (TdT; e.g., 72,033 from Affymetrix, *see* **Note 5**).
- Mineral oil (e.g., M1028 from Sigma).
- Tailing-wash buffer: 50 mM KH_2PO_4 pH 7, 1 mM DTT, 0.25 IGEPAL CA-630. Store at −20 °C.
- Expand Long Template PCR System (ELTPS) kit (Roche) containing Buffer 1 and PolMix. Stored at −20 °C (*see* **Note 6**).
- Formamide, 20%. Stored at −20 °C.
- Oligonucleotide CP2: 5′- TCA GAA TTC ATG CCC CCC CCC CCC CCC-3′. 24 mM dilution Store at −20 °C (*see* **Note 7**).

2.1.2 Plasticware

Plasticware in contact with sample should have low nucleic acid binding properties.

- 0.2 mL domed cap microtubes.
- Microtubes (e.g., 0.5, 1.5, or 2.0 mL) for preparing reagent mixes.
- 10 μl, 20 μl, 200 μl, and 1000 μl filter-tips.
- 50 mL screw-cap tube.

2.1.3 Equipment

- Laminar flow bench.
- Microtube racks.
- Microtube ice racks (optional).
- 0.5–2 μl, 10 μl, 2–20 μl, 20–200 μl, 200–1000 μl micropipettes (recommended).
- Vortexer.
- Benchtop microcentrifuge for microtubes listed above.
- Thermal cycler.
- Tube roller.
- Hybridization oven (*see* **Note 8**).
- Magnetic racks (e.g., DynaMag-96 Side Magnet from Thermo).

2.1.4 Cycling Programs

Ideally, dedicated thermal cycler used only for WTA and WGA should be used, and it should be in the same room. Cycling programs are shown in Table 1.

2.1.5 Reagent Mixes

For buffers, final concentrations of individual components are given. For reagent mixes, volumes of individual reagents sufficient for one sample are given; it is recommended to prepare 5–10% more of the total reagent mix volume than needed for desired number of samples, to account for pipette errors and "dead" volumes. Reagent mixes composition are given in Table 2.

2.2 DNA Precipitation and Whole Genome Amplification

Ideally, this procedure should be performed in the same room and using the same equipment as for the WTA (*see* Subheading 2.1). During execution of the protocol, gloves should be exchanged frequently, especially upon touching anything that is outside of the bench (face, chair, drawer handles, etc.). It is advised not to wear wrist watches, rings, bracelets, etc.

2.2.1 Reagents and Kits

All reagents should be molecular biology grade, and DNase and RNAse free. All dilutions and mixtures should be made with nuclease-free water. Reagent dilutions made from powder should be filtered using 0.22 μm filter. All dilutions should be aliquoted to avoid frequent freeze-thaw cycles.

- Nuclease-free water.
- 70% ethanol.
- WGA buffer pH 7.5: 100 mM Tris-acetate pH 7.8, 100 mM Mg $(CH_3COO)_2$, 500 mM CH_3COOK (*see* **Note 9**). Store at −20 °C.
- 10% Tween 20. Store at −20 °C.
- 10% IGEPAL CA-630. Store at −20 °C.

Table 1
Cycling programs used in Subheading 3.1

Steps	Time/temperature	Repetition
LYSIS		
Step 1	10 min/45 °C	1×
Step 2	1 min/74 °C	1×
Step 3	10 min/22 °C	1×
Step 4	∞/22 °C	
DENATURATION		
Step 1	4 min/94 °C	1×
Step 2	∞/94 °C	1×
TAILING		
Step 1	60 min/37 °C	1×
Step 2	∞/22 °C	1×
TINACTIVATE		
Step 1	5 min/70 °C	1×
Step 2	∞/4 °C	
AMPLIFICATION		
Step 1	30 s/78 °C	1×
Step 2 Step 3 Step 4	15 s/94 °C 30 s 65 °C 2 min/68 °C	20×
Step 5 Step 6 Step 7	15 s/94 °C 30 s/65 °C 2 min 30 s/68 °C +10 s/cycle	20×
Step 8 Step 9 Step 10 Step 11	15 s/94 °C 30 s/65 °C 7 min/68 °C ∞/4 °C	1×

∞—forever

- 10 mg/mL Proteinase K (e.g., 3115828001 from Roche). Store at −20 °C.
- MseI, 50 U/μl (e.g., R0525M from NEB; *see* **Note 10**) Store at −20 °C.
- ddMse11 oligonucleotide: 5′-TAA CTG ACAG ddC-3′. 100 μM. Store at −20 °C.
- Lib1 oligonucleotide: 5′-AGT GGG ATT CCT GCT GTC AGT-3′. 100 μM. Store at −20 °C.

Table 2
Composition of reagent mixes used in Subheading 3.1

Reagent	Volume [μl]
RT-mix 1	
5× SSIIRT buffer	2.0
0.1 M DTT	1.0
10% IGEPAL CA-630	0.5
Nuclease-free water	0.5
100 μM T24 oligonucleotide	3.0
200 μM N8 oligonucleotide	3.0
RT-mix 2	
5× SSIIRT buffer	2.0
0.1 M DTT	1.0
Nuclease-free water	5.0
dNTPs (10 mM each)	1.0
Add later (*see* **step 21** of Subheading 3.1)	
SSII RT enzyme	1.0
Tailing-mix	
40 mM $MgCl_2$	1.0
1 mM DTT	1.0
2 mM dGTP	1.0
0.2 KH_2PO_4	0.5
Nuclease-free water	6.5
PCR-mix 1	
ELTPS buffer 1	4.0
20% formamide	7.5
Nuclease-free water	24.0
PCR-mix 2	
24 μM CP2 oligonucleotide	2.5
dNTPs (10 mM each)	1.75
Add later (*see* **step 39** of subheading 3.1)	
ELTPS PolMix enzyme	1.5

- 10 mM ATP (e.g., 11140965001 from Roche). Store at −20 °C.
- T4 DNA ligase, 5 U/μl (e.g., 10799009001 from Roche). Store at −20 °C.
- dNTP mix, 10 mM each. Store at −20 °C.
- Expand Long Template PCR System (ELTPS) kit (Roche) containing Buffer 1 and PolMix. Store at −20 °C.

2.2.2 Plasticware

Plasticware in contact with sample should have low nucleic acid binding properties.

- 10 μl, 20 μl, 200 μl, and 1000 μl filter-tips.
- 0.2 mL domed cap microtubes.
- Microtubes (e.g., 0.5, 1.5, or 2.0 mL) for preparing reagent mixes.

2.2.3 Equipment

- Laminar flow bench.
- Microtube racks.
- 0.5–2 μl, 10 μl, 2–20 μl, 20–200 μl, 200–1000 μl micropipettes (recommended).
- High speed microcentrifuge with controlled temperature.
- Thermomixer.
- Thermal cycler.

2.2.4 Reagent Mixes

For reagent mixes, volumes of individual reagents sufficient for one sample are given; it is recommended to prepare 5–10% more of the total reagent mix volume than needed for desired number of samples, to account for pipette errors and "dead" volumes. Reagent mixes composition are given in Table 3.

2.2.5 Cycling Programs

Ideally, dedicated thermal cycler used only for WTA and WGA should be used, and it should be in the same room. Cycling programs are shown in Table 4.

2.3 Control of the WTA Quality

This procedure should not be performed in the room where WTA or WGA takes place, nor the same equipment, plasticware, or reagents should be used. This procedure can be performed under the same conditions as any other conventional PCR in your lab.

2.3.1 Reagents and Kits

All reagents should be molecular biology grade, and DNase and RNAse free. All dilutions and mixtures should be made with nuclease-free water. Reagent dilutions made from powder should be filtered using 0.22 μm filter. All dilutions should be aliquoted to avoid frequent freeze-thaw cycles.

Table 3
Composition of reagent mixes used in Subheading 3.2

Reagent	Volume [μl]
Proteinase K digestion mix	
WGA buffer	0.5
10% tween 20	0.13
10% IGEPAL CA-630	0.13
10 mg/mL Proteinase K	0.26
MseI digestion mix	
MseI (50 U/μl)	0.25
Nuclease-free water	0.25
Adapter annealing mix	
WGA buffer	0.5
100 μM Lib1 primer	0.5
100 μM ddMse11 primer	0.5
Nuclease-free water	1.5
Ligation mix	
Adapter annealing mix	3.0
10 mM ATP	1.0
T4 DNA ligase (5 U/μl)	1.0
PCR-mix	
ELTPS buffer 1	3.0
dNTPs (10 mM each)	2.0
ELTPS PolMix enzyme	1.0
Nuclease-free water	34.0

- Taq polymerase with appropriate buffer (e.g., FastStart kit 04738381001 from Roche, containing FastStart Taq polymerase, PCR buffer with 20 mM $MgCl_2$, and dNTP mix). Store at −20 °C.
- BSA, 20 mg/mL. Store at −20 °C.
- dNTP mix, 10 mM each (optional, in case it is not contained in Taq polymerase kit). Store at −20 °C.
- Nuclease-free water. Store at −20 °C.
- *GAPDH* forward primer: 5′- CCA TCT TCC AGG AGC GAG AT-3′, 100 μM. Store at −20 °C.

Table 4
Cycling programs used in Subheading 3.2

Steps	Time/temperature	Repetition
PROTEINASE		
Step 1	15 h/42 °C	1×
Step 2	10 min/80 °C	1×
Step 3	∞/4 °C	1×
MSE1		
Step 1	3 h/37 °C	1×
Step 2	5 min/65 °C	1×
Step 3	∞/4 °C	1×
ANNEAL		
Step 1	1 min/65 °C −1 °C/cycle	50×
Step 2	∞/15 °C	1×
LIGATION		
Step 1	Overnight/15 °C (*see* **Note 11**)	1×
Step 2	∞/15 °C	
AMPLIFICATION		
Step 1	3 min/68 °C	1×
Step 2 Step 3 Step 4	40 s/94 °C 30 s/57 °C 1 min 30 s/68 °C +1 s/cycle	15×
Step 5 Step 6 Step 7	40 s/94 °C 30 s/57 °C +1 °C/cycle 1 min 45 s/68 °C +1 s/cycle	9×
Step 8 Step 9 Step 10	40 s/94 °C 30 s/65 °C 1 min 53 s/68 °C +1 s/cycle	23×
Step 11 Step 12	3 min 40 s/68 °C ∞/4 °C	1×

∞—forever

- *GAPDH* reverse primer: 5′- CAG TGG GGA CAC GGA AGG-3′, 100 μM. Store at −20 °C.
- *ACTB* forward primer: 5′- GCG TGA CAT TAA GGA GAA GCT G-3′, 100 μM. Store at −20 °C.
- *ACTB* reverse primer: 5′- CGC TCA GGA GGA GCA ATG AT-3′, 100 μM. Store at −20 °C.
- *EEF1A1* forward primer: 5′- CTG TGT CGG GGT TGT AGC CA-3′, 100 μM. Store at −20 °C.
- *EEF1A1* reverse primer: 5′- TGC CCC AGG ACA CAG AGA CT-3′, 100 μM. Store at −20 °C.

2.3.2 Plasticware

Plasticware in contact with sample should have low nucleic acid binding properties.

- 0.2 mL microtubes.
- Microtubes (e.g., 0.5, 1.5, or 2.0 mL) for preparing reagent mixes.
- 10 μl, 20 μl, 200 μl, and 1000 μl filter-tips.

2.3.3 Equipment

- PCR bench.
- Microtube racks.
- Microtube ice racks (optional).
- 0.5–2 μl, 10 μl, 2–20 μl, 20–200 μl, 200–1000 μl micropipettes (recommended).
- Benchtop microcentrifuge for microtubes listed above.
- Thermal cycler.

2.3.4 Reagent Mixes

For reagent mixes, volumes of individual reagents sufficient for one sample are given; it is recommended to prepare 5–10% more of the total reagent mix volume than needed for desired number of samples, to account for pipette errors and "dead" volumes. Reagent mixes composition are given in Table 5.

2.3.5 Cycling Programs

Cycling programs are shown in Table 6.

2.4 Control of the WGA Quality

This procedure should not be performed in the room where WTA or WGA takes place, nor the same equipment, plasticware, or reagents should be used. This procedure can be performed under the same conditions as any other conventional PCR in your lab.

2.4.1 Reagents and Kits

All reagents should be molecular biology grade, and DNase and RNAse free. All dilutions and mixtures should be made with nuclease-free water. Reagent dilutions made from powder should be filtered using 0.22 μm filter. All dilutions should be aliquoted to avoid frequent freeze-thaw cycles.

Table 5
Composition of reagent mixes used in Subheading 3.3

Reagent	Volume [μl]
Primer-mix	
Each primer	*8 (6 × 8 μl = 48 μl)*
Nuclease-free water	*152*
PCR-mix	
10× FastStart PCR buffer (with 20 mM $MgCl_2$)	1.0
Primer-mix	1.0
20 mg/mL BSA	0.2
dNTPs (10 mM each)	0.2
FastStart Taq polymerase	0.1
Nuclease-free water	6.5

Table 6
Cycling programs used in Subheading 3.3

Steps	Time/temperature	Repetition
WTAQC		
Step 1	4 min/95 °C	1×
Step 2 Step 3 Step 4	30 s/95 °C 30 s/58 °C 1 min 30 s/72 °C	32×
Step 5 Step 6	7 min/72 °C ∞/4 °C	1×

∞—forever

- Taq polymerase with appropriate buffer (e.g., FastStart kit 04738381001 from Roche, containing FastStart Taq polymerase, PCR buffer with 20 mM $MgCl_2$, and dNTP mix). Store at −20 °C.
- 20 mg/mL BSA. Store at −20 °C.
- dNTP mix, 10 mM each (optional, in case it is not contained in Taq polymerase kit). Store at −20 °C.
- Nuclease-free water. Store at −20 °C.

- *KRAS* Forward primer: 5′-ATA AGG CCT GCT GAA AAT GAC-3′, 100 μM. Store at −20 °C.
- *KRAS* Reverse primer: 5′-CTG AAT TAG CTG TAT CGT CAA GG-3′, 100 μM. Store at −20 °C.
- D5S2117 Forward primer: 5′-CCA GGT GAG AAC CTA GTC AG-3′, 100 μM. Store at −20 °C.
- D5S2117 Reverse primer: 5′-ACT GTG TCC TCC AAC CAT GG-3′, 100 μM. Store at −20 °C.
- *KRT19P1* Forward primer: 5′-GAA GAT CCG CGA CTG GTA C-3′, 100 μM. Store at −20 °C.
- *KRT19P1* Reverse primer: 5′-TTC ATG CTC AGC TGT GAC TG-3′, 100 μM. Store at −20 °C.
- *TP53* Forward primer: 5′-GAA GCG TCT CAT GCT GGA TC-3′, 100 μM. Store at −20 °C.
- *TP53* Reverse primer: 5′-CAG CCC AAC CCT TGT CCT TA-3′, 100 μM. Store at −20 °C.

2.4.2 Plasticware

Plasticware in contact with sample should have low nucleic acid binding properties.

- 0.2 mL microtubes.
- Microtubes (e.g., 0.5, 1.5, or 2.0 mL) for preparing reagent mixes.
- 10 μl, 20 μl, 200 μl, and 1000 μl filter-tips.

2.4.3 Equipment

- PCR bench.
- Microtube racks.
- Microtube ice racks (optional).
- 0.5–2 μl, 10 μl, 2–20 μl, 20–200 μl, 200–1000 μl micropipettes (recommended).
- Benchtop microcentrifuge for microtubes listed above.
- Thermal cycler.

2.4.4 Reagent Mixes

For reagent mixes, volumes of individual reagents sufficient for one sample are given; it is recommended to prepare 5–10% more of the total reagent mix volume than needed for desired number of samples, to account for pipette errors and "dead" volumes. Reagent mixes composition are given in Table 7.

2.4.5 Cycling Programs

Cycling programs are shown in Table 8.

Table 7
Composition of reagent mixes used in Subheading 3.4

Reagent	Volume [μl]
Primer-mix	
Each primer	*8 (8 × 8 μl = 64 μl)*
Nuclease-free water	*136*
PCR-mix	
10× FastStart PCR buffer (with 20 mM $MgCl_2$)	1.0
Primer-mix	1.0
20 mg/mL BSA	0.2
dNTPs (10 mM each)	0.2
FastStart Taq polymerase	0.1
Nuclease-free water	6.5

Table 8
Cycling programs used in Subheading 3.4

Steps	Time/temperature	Repetition
WGAQC		
Step 1	4 min/95 °C	1×
Step 2 Step 3 Step 4	30 s/95 °C 30 s/58 °C 1 min 30 s/72 °C	32×
Step 5	7 min/72 °C	1×
Step 6	∞/4 °C	

∞—forever

3 Methods

3.1 Isolation of mRNA and Genomic DNA, and Whole Transcriptome Amplification

The starting point is the sample in 0.2 mL domed cap microtubes containing one or more cells in 4.4 μl of mTRAP lysis buffer supplemented with 10 ng of *E. coli* tRNA (*see* **Note 12**). The cells can be freshly isolated or kept at −80 °C.

1. Before bringing samples, prepare mixture containing mTRAP lysis buffer and mTRAP protease mixed in 20:1 ratio. Take 1 μl of this mixture and mix with 1 μl of mTRAP biotinylated poly-T PNA oligonucleotides (*see* **Notes 13, 14**, and **15**).

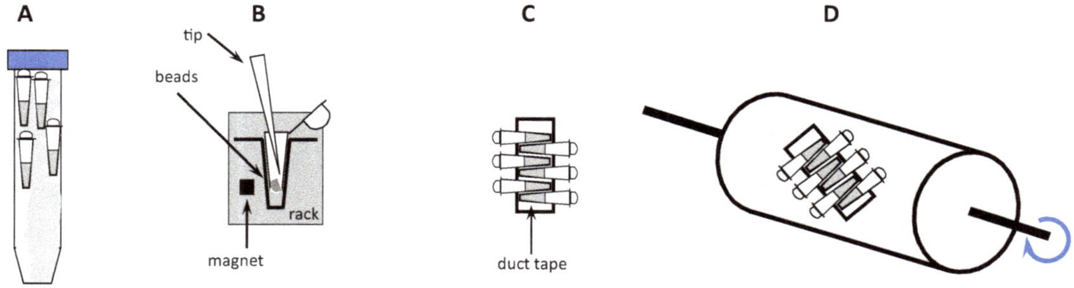

Fig. 2 Important steps during WTA. (**a**) Microtubes in 50 mL screw cap tube stuffed with paper wipes (grey). (**b**) Position of the pipette tip during bead washing and supernatant removal. (**c**) Fastening the microtubes on the stripe of the masking duct tape. (**d**) Fastening the microtubes using masking duct tape to the vial/bottle of hybridization oven

2. Bring samples into the laminar flow bench and keep on ice (*see* **Note 16**).
3. Add 2 μl of lysis buffer:protease/PNA mix into the sample (*see* **Notes 17** and **18**). Shortly spin in microcentrifuge.
4. Lyse the sample in thermal cycler, using cycling program LYSIS (Table 1). *See* **Note 19**.
5. Prepare and label two sets of fresh 0.2 mL domed cap microtubes. One set will be used for downstream WTA procedure and the other for WGA procedure (*see* Subheading 3.2). *See* **Note 20**.
6. Add 0.8 μl of polyacrylamide carrier into the WGA-set of 0.2 mL tubes from previous step (*see* **Note 21**). Put these microtubes aside.
7. Thoroughly vortex mTRAP streptavidin-conjugated microbeads and shortly spin in microcentrifuge (*see* **Note 22**).
8. Once the lysis is done, bring the samples under the laminar flow bench, shortly spin them, place them on ice, and add 4 μl of beads into each sample.
9. Stuff some paper towels into 50 mL tube, and put 0.2 microtubes with samples in. Make sure that the microtubes are not moving (Fig. 2a). Place the 50 mL tube onto the roller and incubate 45 min at room temperature (*see* **Notes 23** and **24**).
10. Shortly before the incubation is done, turn on the hybridization oven and set it to 44 °C.
11. Prepare RT-Mix 1 and RT-Mix 2 (without RT enzyme) and keep them on ice (Table 2). *See* **Note 25**.
12. Once the incubation is over, spin the microtubes and place them on ice.

13. Add 10 µl of IGEPAL-wash buffer into each tube, gently mix by pipetting up and down, and place on magnetic microtube rack kept on ice (*see* **Notes 17** and **26**). Place the microtubes in a row where the magnet is on the same side of the well where the hand is that you are using to pipet (Fig. 2b; *see* **Note 27**).
14. Carefully, placing the pipette tip on away from the beads, take out 20 µl of supernatant and transfer it to WGA-microtubes with polyacrylamide carrier (*see* **Notes 28** and **29**).
15. Add 20 µl of Tween-wash buffer into each tube and gently mix by pipetting up and down. Collect the compete sample into the pipette tip and transfer to a fresh WTA microtube (from **step 5**) and place it on magnetic microtube rack kept on ice (*see* **Note 30**).
16. Repeat **step 14**.
17. Add 20 µl of IGEPAL-wash buffer into each tube, gently mix by pipetting up and down, and place on magnet microtube rack kept on ice.
18. Carefully, placing the pipette tip on away from the beads, take out all of the supernatant (slightly more than 20 µl) and transfer it to WGA-microtubes with polyacrylamide carrier (*see* **Notes 31** and **32**).
19. Add 10 µl of the RT-Mix 1 to the beads, mix by pipetting up and down, and incubate 10 min at room temperature (in a non-magnetic microtube rack; *see* **Note 33**).
20. Add 120 µl of 100% ice-cold ethanol to collected supernatants in WGA-microtubes. Mix by inverting microtubes (*see* **Note 34**). Store the microtubes at −20 °C until processing (*see* Subheading 3.2).
21. Add 1 µl of SSII RT enzyme to RT-mix 2. Mix well.
22. Add 10 µl of RT Mix 2 to each sample.
23. Fasten the microtubes to the outer side of the hybridization vial and place inside the pre-heated hybridization oven (Fig. 2c, d; *see* **Note 35**). Start the rotation and incubate for 45 min at 44 °C (*see* **Notes 8** and **36**).
24. After incubation, shortly spin the microtubes and place them on magnetic rack kept on ice.
25. Prepare Tailing-mix (Table 2).
26. Carefully, placing the pipette tip away from the beads, take out 20 µl of supernatant and discard (Fig. 2b).
27. Add 20 µl of Tailing-wash buffer, gently mix by pipetting up and down, and place on magnetic microtube rack kept on ice.
28. Carefully, placing the pipette tip away from the beads, take out 20 µl of supernatant and discard (*see* **Note 37**).

29. Add 10 μl of Tailing-mix.
30. Add 40 μl of oil (*see* **Note 38**).
31. Start DENATURATION (Table 1) program on the thermal cycler and wait until the block temperature starts to rise. Pause the program, and wait until the block temperature reaches 94 °C. Place the microtubes inside, close the lid, and restart the program (*see* **Note 39**).
32. After 4 min open the lid, take the microtubes out, and place them on ice.
33. Carefully, add 0.8 μl of TdT enzyme underneath the oil.
34. Put the microtubes in the thermal cycler and start the program TAILING (Table 1; *see* **Notes 40** and **41**).
35. After tailing, inactivate TdT by starting TINACTIVATE program on thermal cycler (Table 1).
36. Prepare PCR-mix 1 and PCR-Mix 2 (without PolMix enzyme; Table 2).
37. Spin the microtubes and place them on ice.
38. Add 35.5 μl of PCR-mix 1 into each sample (*see* **Note 42**).
39. Add 1.5 μl of PolMix enzyme into PCR-Mix 2.
40. Place the microtubes in thermal cycler and start AMPLIFICATION program (Table 1).
41. Wait until the block temperature reaches 78 °C and pause the program.
42. Quickly add 5.5 μl of PCR-mix 2 into each sample (*see* **Note 43**).
43. Close the lid of the thermal cycler and restart the program (*see* **Notes 44** and **45**).

3.2 DNA Precipitation and Whole Genome Amplification

Supernatants collected during WTA procedure (**step 20** of Subheading 3.1) are stored at −20 °C before processing. The procedure described in this section takes 4 consecutive days. Breaks in terms of skipping days are possible, but are strongly discouraged and should be used only in case of emergency (*see* **Note 46**).

1. Cool the centrifuge to 4 °C.
2. Centrifuge the DNA-containing supernatants in ethanol 45 min at 18400 rcf at 4 °C (*see* **Note 47**).
3. After centrifuging small white pellets should be visible at the bottom of the tube (*see* **Note 48**). Total volume in the microtube is around 180 μl.
4. Remove and discard supernatant, and leave approximately 20 μl inside (*see* **Notes 49** and **50**).
5. Gently add 180 μl of 70% ethanol (*see* **Note 51**).

6. Place the tubes in Thermomixer and incubate 10 min on 19 °C at 350 rpm (*see* **Note 52**).
7. Set the centrifuge to room temperature.
8. After incubation, centrifuge the samples 10 min at 18400 rcf at room temperature.
9. Repeat **steps 4–8** two more times, for a total of three ethanol washing/thermomixer/centrifuging steps.
10. After the final centrifugation, remove and discard complete supernatant (*see* **Note 53**).
11. Leave microtubes open for the pellets to dry. This may take between few minutes, e.g., 30 min (*see* **Notes 54** and **55**).
12. Once the pellets are dry, add 3.48 μl of water (*see* **Note 56**).
13. Incubate overnight in Thermomixer on 19 °C at 350 rpm.
14. On the next day, take the samples out of Thermomixer, shortly spin, and place on microtube rack.
15. Prepare Proteinase-K digestion mix (Table 3). Add 1.02 μl to each sample. Incubate overnight in thermal cycler using program PROTEINASE (Table 4). *See* **Notes 57** and **58**.
16. On the following day, prepare MseI digestion mix (Table 3). Add 3 μl to each sample, and incubate in thermal cycler using program MSE1 (Table 4). *See* **Note 59**.
17. Prepare Adapter annealing mix (Table 3). Incubate the complete reagent mix in thermal cycler using program ANNEAL (Table 4). *See* **Note 60**.
18. After MseI digestion and adapter annealing are finished, prepare Ligation mix (Table 3). Add 5 μl of the Ligation mix to each sample, and incubate overnight in thermal cycler using program LIGATION (Table 4). *See* **Note 61**.
19. On the final day, prepare PCR-mix (Table 3). Add 40 μl to each sample and run the program AMPLIFICATION (Table 4) in thermal cycler. *See* **Notes 62, 63**, and **64**.

3.3 Checking the Quality of WTA

This procedure is a multiplex end-point PCR that amplifies three fragments of different lengths, mapping to three "housekeeping" genes (*see* **Note 65**).

1. Label enough 0.2 microtubes for WTA samples, positive, and negative control (*see* **Note 66**).
2. Prepare the PCR-mix (Table 5) and distribute 9.0 μl of the mix to the tubes.
3. Add 1 μl of the WTA product. Shortly spin the tubes, place them in thermal cycler, and run program WTAQC (Table 6).

4. After amplification, perform the fragment analysis using your preferred way (*see* **Note 67**).
5. Note down the number of expected amplified fragments. The number of bands reflects the quality of the WTA sample (*see* **Notes 68** and **69**).

3.4 Checking the Quality of WGA

This procedure is a multiplex end-point PCR that amplifies four fragments of different lengths, located on MseI digested fragments of different lengths and mapping to different regions of the genome (*see* **Note 70**).

1. Label enough 0.2 microtubes for WGA samples, positive, and negative control (*see* **Note 71**).
2. Prepare the PCR-mix (Table 7) and distribute 9.0 μl of the mix to the tubes.
3. Add 1 μl of the WGA product. Shortly spin the tubes, place them in thermal cycler, and run program WGAQC (Table 8).
4. After amplification, perform the fragment analysis using your preferred way (*see* **Note 67**).
5. Note down the number of expected amplified fragments. The number of bands reflects the quality of the WGA sample (*see* **Note 72**).

4 Notes

1. Alternatively, linear polyacrylamide can be prepared using published protocol [19].
2. This is an established protocol that uses SuperScript II reverse transcriptase. It is possible to use other SuperScript enzymes (e.g., III or IV) or enzymes from other manufacturers, as long as they have optimal processivity at 44 °C and are RNaseH-deficient. Replacement of an RT system can be done with sufficient comparison and testing using SS II system.
3. N8 oligonucleotide contains XbaI restriction site that facilitates removal of oligonucleotide adapters, cloning of individual fragments, or preparation of sequencing libraries. The sequence of this oligonucleotide can be modified by introducing or adding other restriction sites.
4. T24 oligonucleotide contains XbaI and BpuEI restriction sites that facilitate removal of oligonucleotide adapters, cloning of individual fragments, or preparation of sequencing libraries. The sequence of this oligonucleotide can be modified by introducing or adding other restriction sites; e.g., previous version of this oligonucleotide contained only XbaI site (5′-CCC CCC CCC CCC CCC GTC TAG ATT TTT TTT TTT TTT TTT TTT TTT TVN-3′).

5. Other TdT enzymes can be used, but they have to be compatible with the Tailing reaction mix used here (Table 2); i.e., own TdT reaction buffer cannot be used.
6. Other polymerase blends can be used. Prerequisites are that the blend has to generate long DNA fragments and that it has proofreading activity. For some downstream applications, it may be necessary that the composition of the reaction buffer is known.
7. CP2 oligonucleotide contains EcoRI restriction sites that facilitate removal of oligonucleotide adapters, cloning of individual fragments, or preparation of sequencing libraries. The sequence of this oligonucleotide can be modified by introducing or adding other restriction sites.
8. Any alternative approach that enables fixing the tubes to a rotating element at 44 °C is also possible.
9. pH of the WGA buffer is set the following way. After mixing of all components in a large volume, an aliquot is taken and pH measured. Then, using NaOH or HCl, pH is further set. Again, an aliquot is measured on pH meter. And so on. This is done to avoid possible contamination of the buffer by general pH meter. After the pH is finally set, the buffer should be filtered, aliquoted, and frozen.
10. Alternative MseI enzyme can be used, as long as it has the same enzyme concentration and it works with WGA buffer.
11. The "overnight" ligation step does not have a fixed amount of time. 12 h should be enough. This offers some flexibility in the execution of the different steps of WGA protocol.
12. *E. coli* tRNA serves two purposes. First, it coats the walls of the microtube, thereby preventing passive binding of precious sample's RNA. Second, since it is added in excess, it serves as a substrate for any RNases that may be active before the sample is processed. However, if addition of tRNA is not possible or is not practical, it can be left out. This protocol really starts after the cell isolation step.
13. All reagents should be thawed, thoroughly mixed, and spun in microcentrifuge before preparing mixes. Upon adding reagents or mixes to the samples, unless otherwise noted, mixing by gently pipetting up and down and a short spin in a microcentrifuge should always be done. Additional spin should also be done after any incubation steps, and before adding new reagents/mixes.
14. It is recommended not to process more than 12 samples at once (including controls). Each run should have at least one "cell-free/reagents only" negative control. Optionally, a positive control containing one or more cells of, e.g., cell line or peripheral blood lymphocytes (to name the few that can be easily obtained) can be included.

15. Depending on the number of samples, the simplest approach is to mix 9.5 μl of lysis buffer with 0.5 μl of protease, or 19 μl of lysis buffer with 1 μl of protease. Then, another mix is made by mixing equal volumes of lysis buffer/protease mix and PNA oligonucleotides (e.g., 13 μl + 13 μl for 12 samples).
16. Microtube racks with samples should always be kept on ice, unless otherwise noted. Instead of ice, one can use microtube ice racks.
17. Pipetting steps should be done by most appropriate pipette and tip for the volume being handled. That being said, it is better to pipet, for example, 2 μl with a 0.5–2 μl pipet than 2–20 μl, and it is better to pipet 20 μl with a 2–20 μl pipet than 20–200 μl pipet.
18. The reagent mix should be added by touching the microtube wall near the liquid surface. Touching of the sample should be avoided due to capillary forces that may pull in minuscule amounts of liquid that may result in loss of nucleic acid molecules.
19. During this step, cell is lysed, enzymes thermally inactivated and digested, and biotinylated poly-T PNAs capture poly-A tails of mRNAs (Fig. 1a).
20. This protocol can be used for WTA only. If genomic DNA for WGA is not needed, the second set of 0.2 mL microtubes with polyacrylamide carrier is not needed.
21. Polyacrylamide carrier serves to facilitate and visualize DNA precipitation.
22. mTRAP beads should be vortexed or mixed until no bead precipitate is visible and the liquid becomes homogeneous. Short spin in microcentrifuge is needed to remove the liquid from the cap, but it should lead to repeated precipitation of beads. If pipetting takes too long, it should be checked whether the beads started to precipitate, which may require another mixing/vortexing.
23. Any alternative approach that enables fixing the tubes to a rotating element on room temperature is also possible.
24. After this step, biotinylated poly-T PNAs hybridized to poly-A tails of mRNAs are captured by streptavidin-coated beads (Fig. 1a). From this step onward, all reactions take place on beads, i.e., on solid phase.
25. IGEPAL prevents aggregation of beads. T24 oligonucleotide hybridizes to the beginning of the poly-A tail of mRNA (not hybridized by PNA oligonucleotide). N8 is similar, but instead of poly-T and following dinucleotide sequence, it contains random octamer, to bind randomly along the mRNA, thus creating short cDNA fragments. Octamer is used instead of

commonly used hexamers to reduce the number of binding sites and thereby increase the length of cDNA fragments.

26. Detergents reduce the surface tension and facilitate supernatant removal.
27. If you are right-handed, the beads will precipitate on the right wall of the microtube, and pipette tip can be placed on the opposite side to collect the supernatant.
28. If genomic DNA is not needed, the supernatant can be discarded.
29. The washing of the beads is the critical step and should be done very carefully and slowly. When adding wash buffers, bubbles should be avoided. If they appear, be very careful in removing the supernatant, since the bubbles pull the beads into the pipette tip. In that case, it is better to remove less than 20 μl. One can remove supernatant from a microtube and immediately proceed with the next washing step and put back the microtube to the magnetic rack. Alternatively, supernatant can be sequentially removed from up to 4 microtubes (close the lid of the microtubes where the supernatant is removed) and then new wash can be performed. In any case, it is critically important not to let the beads to dry.
30. This is the earliest safest point to get rid of the original sample microtube, thereby reducing the risk to carry-over any unwanted residues.
31. To ensure that complete supernatant is removed, one approach can be to press the pipette plunger slightly into the "second" step before inserting the tip into the sample, thereby being able to collect slightly more than 20 μl.
32. If mRNAs and WTA are not needed, the sample with the beads can be discarded, and supernatant in WGA microtubes can be processed further, as described in Subheading 3.2. However, if only genomic DNA and WGA of single cell are needed, there are faster, simpler, and cheaper protocols [20, 21].
33. During this step, T24 and N8 oligonucleotides are binding to mRNAs (Fig. 1a).
34. This step will enable precipitation of DNA.
35. For this we usually stick the microtubes to a strip of paper/masking duct tape, and fasten it to the vial/bottle of the hybridization oven (Fig. 2c, d).
36. During this step, reverse transcription takes place (Fig. 1a).
37. These steps of beads washing and supernatant removal are critical. The purpose is to remove any remaining dNTPs from

the RT step, since they may interfere with subsequent poly-G tailing (by introducing dNTPs other than dGTP).

38. Oil serves to prevent evaporation of the sample during subsequent denaturation step.
39. During the denaturation step, mRNAs and cDNAs are released from beads. From this step onward, the reactions are not happening on the solid phase.
40. In this step, TdT will incorporate dGTPs and will generate 3′ poly-G tails on single-stranded cDNAs (Fig. 1a).
41. Although not recommended, this is a step where it is possible to introduce a break, e.g., overnight. One can simply leave the samples in the cycler or at 22 °C.
42. Formamide is used to decrease the annealing temperature of G-C pairs formed by CP2 oligonucleotides. CP2 oligonucleotides will bind to the poly-G tails generated in **step 34** of Subheading 3.1. The same nucleotide will bind to the poly-G sequences complementary to the 5′ end of the T24 and N8 oligonucleotides introduced during reverse transcription step. Therefore, CP2 oligonucleotide serves both as forward and reverse primer (Fig. 1a).
43. This is basically a "hot-start" approach. The reason for the hot start is to avoid unspecific priming and extension of the CP2 primers (that bound to the single-stranded cDNA at low temperatures) until 94 °C is reached.
44. This is a final step of a global mRNA amplification (i.e., WTA).
45. Theoretically, original WTA product can perpetually be used to reamplify itself. PCR mix can be prepared according to Table 9 and, with 1 μl of WTA product added, can be amplified using cycling program in Table 10. Quality can be controlled by using the procedure shown in Subheading 3.3.
46. There is a kit based on WGA protocol presented here. This kit (Ampli1 WGA PLUS Kit) is marketed by Menarini Silicon Biosystems, and its use is previously published 20 and also described elsewhere in this book.
47. Centrifuging time can also be a bit longer, e.g., 60 min.
48. Pellets may also have the shades of brown, coming from beads that were collected during supernatant removal (**steps 13–18** of Subheading 3.1). These beads do not interfere with the procedure.
49. Pipet very slowly. Avoid detaching of the pellet from the bottom of the microtube, touching the pellet by the tip, or pipetting near the pellet.
50. If the pellet is detached, it is not a big problem, just be very cautious not to suck it in in the tip. If it does happen, expel it

Table 9
PCR-mix for WTA reamplification

Reagent	Volume [μl]
PCR-mix	
ELTPS buffer 1	5.0
24 μM CP2 oligonucleotide	6.0
dNTPs (10 mM each)	1.75
20% Formamide	7.5
ELTPS PolMix enzyme	1.5
Nuclease-free water	27.25

Table 10
Cycling profile for WTA reamplification

Steps	Time/temperature	Repetition
WTAQC		
Step 1	1 min/95 °C	1×
Step 2 Step 3 Step 4	15 s/94 °C 1 min/55 °C 3 min 30 s/65 °C	5×
Step 5 Step 6 Step 7	15 s/94 °C 1 min/55 °C 3 min 30 s/65 °C +10 s/cycle	3×
Step 8	7 min/65 °C	1×
Step 9	∞/4 °C	

∞—forever

out, and briefly spin the microtube again. If the pellet sticks to the surface of the tips, just pipet vigorously. At this point, damaged pellet is better than the lost pellet. Repeat supernatant removal, as described in **step 4** of Subheading 3.2.

51. Just add ethanol along the microtube wall. Do not mix or resuspend.
52. These steps serve to wash away buffer residues or other contaminating molecules from the DNA pellet.
53. In this step the complete supernatant is removed. Therefore, pipetting near the pellet cannot be avoided. Just be extra slow

and careful. If it is not possible to remove all of the liquid without the danger of sucking in the pellet, rather leave some liquid inside. This will lead to longer drying in **step 11** of Subheading 3.2.

54. White pellets become transparent when dry. Dark brown pellets become light brown when dry. The pellets should not have any visible liquid left (ethanol interferes with subsequent steps), but should also not over-dry, as this may interfere with dissolving in water.
55. Extra caution needed—dry pellets may react to static electricity coming from the hand gloves, and may change location inside of the microtube, or even fly out of the tube, helped by the air flow in the bench. Before taking the tubes to check if the pellet is dry, close the cap, and open it again once the microtube is back in the rack if additional drying is needed.
56. Add water drop near the pellet, close the microtube, and gently flick the tube to enable contact of water and pellet. Do not mix or resuspend.
57. This step may seem superfluous since the protease step was already done during WTA procedure (**Steps 3–4** of Subheading 3.1). However, this step is part of the original WGA protocol and it makes sure that the DNA is completely available for subsequent restriction digestion.
58. From this point on, reagent mixes are just added to the sample by pipetting on the microtube wall and shortly spinning the microtube to bring the mix to the sample. Avoid touching the sample with pipette tip, as this may result in a loss of single-cell DNA.
59. During this step, MseI digest complete genomic DNA of a single cell, thereby generating so called "MseI fragments" (Fig. 1b).
60. During this step, oligonucleotides Lib1 and ddMse11 generate short double-stranded adapters perfectly matching the MseI restriction sites (Fig. 1b).
61. During this step, ligation of adapters takes place. 3′-dideoxy nucleotide of ddMse11 primer prevents extension of this primer and therefore random priming events. Due to the lack of 5′ phospahate, this primer will not be ligated to the 3′ of the MseI fragment overhang and dissociates during the first denaturation step of the global amplification (Fig. 1b). Therefore, only Lib1 primer gets ligated to MseI fragments and forms an identical adapter sequence on both termini of each MseI fragment.
62. Lib1 primer drives the amplification of MseI fragments by acting as both forward and reverse primer (Fig. 1b).

Table 11
PCR-mix for WGA reamplification

Reagent	Volume [μl]
PCR-mix	
ELTPS buffer 1	5.0
10 μM Lib1	5.0
dNTPs (10 mM each)	1.75
20 mg/mL BSA	1.25
ELTPS PolMix enzyme	0.5
Nuclease-free water	35.5

Table 12
Cycling profile for WGA reamplification

Steps	Time/temperature	Repetition
WTAQC		
Step 1	1 min/94 °C	1×
Step 2	30 s/60 °C	1×
Step 3	2 min/68 °C	1×
Step 4 Step 5 Step 6	30 s/94 °C 30 s/60 °C 3 min/68 °C +20 s/cycle	11×
Step 7	∞/4 °C	

∞—forever

63. This is a final step of a global genomic DNA amplification (i.e., WGA).

64. Theoretically, original WGA product can perpetually be used to reamplify itself. PCR mix can be prepared according to Table 11 and, with 1 μl of WGA product added, can be amplified using cycling program in Table 12. Quality can be controlled by using the procedure shown in Subheading 3.4. Please note that elongation can also be performed at 65 °C. This approach generates slightly different fragment distribution. Alternatively, two reamplifications can be done with different elongation temperatures, and resulting WGA reamplification products can be mixed. The utility of each approach should be tested by individual lab according to their downstream applications.

65. *GAPDH* primers generate fragment 489 bp long, *ACTB* primers generate fragment 378 bp long, and *EEF1A1* primers generate fragment 290 bp long.
66. Positive control can be WTA product of a cell pool (cell lines or blood lymphocytes).
67. Standard agarose gel electrophoresis can be used, or devices that use capillary electrophoresis can be used (e.g., Qiaxcel).
68. We have observed that the samples with all three fragments coming in QC-PCR have the best WTA quality. However, these are not absolutes, and certainly samples with 2 or 1 band can be further analyzed by, e.g., sequencing. However, a caution should be exercised in interpretation, since sub-optimal WTA quality may affect both the presence and the amount of transcripts, which may interfere with biological interpretation. Therefore, each application would need a series of validation experiments.
69. Further quality control steps can be undertaken. It is possible to measure the concentration of DNA by using, e.g., Nanodrop or Qubit, or fragment length distribution can be assessed by, e.g., Bioanalyzer. Due to the presence of unbound oligonucleotides, dNTPs, etc., it is advised to purify WTA product before the measurement. This can be done by standard PCR purification kits. In this case, only an aliquot of original WTA product should be used for this, in case anything goes wrong during purification.
70. *KRAS* primers generate fragment 91 bp long, amplifying portion of MseI fragment 192 bp long, originating from chromosome 12. D5S2117 primers generate fragment approx. 140 bp long (this is microsatellite locus and the actual fragment length may vary by few bp or appear as a double band), amplifying portion of MseI fragment 1376 bp long, originating from chromosome 5. *KRT19P1* primers generate fragment 621 bp long, amplifying portion of MseI fragment 1146 bp long, originating from chromosome 12. *TP53* primers generate fragment 301 bp long, amplifying portion of MseI fragment 1374 bp long, originating from chromosome 17.
71. Positive control can be WGA product of a cell pool (cell lines or blood lymphocytes).
72. QC PCR for WGA has 4 PCR fragments. Practice shows that the more fragments are appearing in QC PCR, the WGA product is of better quality. However, these are not absolutes, and certainly samples with less than 4 fragments can be further analyzed by, e.g., sequencing. We advise using GII (genomic integrity index, [21]) to judge the WGA quality. It is considered that WGA products with GII 3 and 4 have good quality, while the WGA product with GII 2 can be used with caution.

References

1. Hay SB, Ferchen K, Chetal K, Grimes HL, Salomonis N (2018) The human cell atlas bone marrow single-cell interactive web portal. Exp Hematol 68:51–61. https://doi.org/10.1016/j.exphem.2018.09.004
2. Gužvić M, Braun B, Ganzer R, Burger M, Nerlich M, Winkler S, Werner-Klein M, Czyz ZT, Polzer B, Klein CA (2014) Combined genome and transcriptome analysis of single disseminated cancer cells from bone marrow of prostate cancer patients reveals unexpected transcriptomes. Cancer Res 74(24): 7383–7394. https://doi.org/10.1158/0008-5472.CAN-14-0934
3. Lee J, Hyeon DY, Hwang D (2020) Single-cell multiomics: technologies and data analysis methods. Exp Mol Med 52(9):1428–1442. https://doi.org/10.1038/s12276-020-0420-2
4. Klein CA, Schmidt-Kittler O, Schardt JA, Pantel K, Speicher MR, Riethmuller G (1999) Comparative genomic hybridization, loss of heterozygosity, and DNA sequence analysis of single cells. Proc Natl Acad Sci U S A 96(8): 4494–4499. https://doi.org/10.1073/pnas.96.8.4494
5. Klein CA, Seidl S, Petat-Dutter K, Offner S, Geigl JB, Schmidt-Kittler O, Wendler N, Passlick B, Huber RM, Schlimok G, Baeuerle PA, Riethmuller G (2002) Combined transcriptome and genome analysis of single micrometastatic cells. Nat Biotechnol 20(4): 387–392. https://doi.org/10.1038/nbt0402-387
6. Klein CA, Zohlnhofer D, Petat-Dutter K, Wendler N (2003) Gene expression analysis of a single or few cells. Curr Protoc Mol Biol Chapter 25:Unit 25B 28. https://doi.org/10.1002/0471142727.mb25b08s61
7. Hahtola S, Burghart E, Jeskanen L, Karenko L, Abdel-Rahman WM, Polzer B, Kajanti M, Peltomaki P, Pettersson T, Klein CA, Ranki A (2008) Clinicopathological characterization and genomic aberrations in subcutaneous panniculitis-like T-cell lymphoma. J Invest Dermatol 128(9):2304–2309. https://doi.org/10.1038/jid.2008.6
8. Chambers KF, Pearson JF, Pellacani D, Aziz N, Gužvić M, Klein CA, Lang SH (2011) Stromal upregulation of lateral epithelial adhesions: gene expression analysis of signalling pathways in prostate epithelium. J Biomed Sci 18(1):45. https://doi.org/10.1186/1423-0127-18-45
9. Hittinger M, Czyz ZT, Huesemann Y, Maneck M, Botteron C, Kaeufl S, Klein CA, Polzer B (2013) Molecular profiling of single Sca-1+/CD34+,- cells--the putative murine lung stem cells. PLoS One 8(12):e83917. https://doi.org/10.1371/journal.pone.0083917
10. Suzuki T, Asami M, Hoffmann M, Lu X, Gužvić M, Klein CA, Perry ACF (2016) Mice produced by mitotic reprogramming of sperm injected into haploid parthenogenotes. Nat Commun 7:12676. https://doi.org/10.1038/ncomms12676
11. Asami M, Lam BYH, Hoffmann M, Suzuki T, Lu X, Yoshida N, Ma MK, Rainbow K, Gužvić M, VerMilyea MD, Yeo GSH, Klein CA, Perry ACF (2023) A program of successive gene expression in mouse one-cell embryos. Cell Rep 42(2):112023. https://doi.org/10.1016/j.celrep.2023.112023
12. Hartmann CH, Klein CA (2006) Gene expression profiling of single cells on large-scale oligonucleotide arrays. Nucleic Acids Res 34(21): e143. https://doi.org/10.1093/nar/gkl740
13. Fuhrmann C, Schmidt-Kittler O, Stoecklein NH, Petat-Dutter K, Vay C, Bockler K, Reinhardt R, Ragg T, Klein CA (2008) High-resolution array comparative genomic hybridization of single micrometastatic tumor cells. Nucleic Acids Res 36(7):e39. https://doi.org/10.1093/nar/gkn101
14. Czyz ZT, Hoffmann M, Schlimok G, Polzer B, Klein CA (2014) Reliable single cell array CGH for clinical samples. PLoS One 9(1):e85907. https://doi.org/10.1371/journal.pone.0085907
15. Hoffmann M, Pasch S, Schamberger T, Maneck M, Mohlendick B, Schumacher S, Brockhoff G, Knoefel WT, Izbicki J, Polzer B, Stoecklein NH, Klein CA (2018) Diagnostic pathology of early systemic cancer: ERBB2 gene amplification in single disseminated cancer cells determines patient survival in operable esophageal cancer. Int J Cancer 142(4): 833–843. https://doi.org/10.1002/ijc.31108
16. Durst FC, Grujovic A, Ganser I, Hoffmann M, Ugocsai P, Klein CA, Czyz ZT (2019) Targeted transcript quantification in single disseminated cancer cells after whole transcriptome amplification. PLoS One 14(8): e0216442. https://doi.org/10.1371/journal.pone.0216442
17. Werner-Klein M, Grujovic A, Irlbeck C, Obradovic M, Hoffmann M, Koerkel-Qu H, Lu X, Treitschke S, Kostler C, Botteron C, Weidele K, Werno C, Polzer B, Kirsch S, Gužvić M, Warfsmann J, Honarnejad K, Czyz Z, Feliciello G, Blochberger I,

Grunewald S, Schneider E, Haunschild G, Patwary N, Guetter S, Huber S, Rack B, Harbeck N, Buchholz S, Rummele P, Heine N, Rose-John S, Klein CA (2020) Interleukin-6 trans-signaling is a candidate mechanism to drive progression of human DCCs during clinical latency. Nat Commun 11(1):4977. https://doi.org/10.1038/s41467-020-18701-4

18. Stoecklein NH, Erbersdobler A, Schmidt-Kittler O, Diebold J, Schardt JA, Izbicki JR, Klein CA (2002) SCOMP is superior to degenerated oligonucleotide primed-polymerase chain reaction for global amplification of minute amounts of DNA from microdissected archival tissue samples. Am J Pathol 161(1): 43–51. https://doi.org/10.1016/S0002-9440(10)64155-7

19. Gaillard C, Strauss F (1990) Ethanol precipitation of DNA with linear polyacrylamide as carrier. Nucleic Acids Res 18(2):378. https://doi.org/10.1093/nar/18.2.378

20. Czyz ZT, Klein CA (2015) Deterministic whole-genome amplification of single cells. Methods Mol Biol 1347:69–86. https://doi.org/10.1007/978-1-4939-2990-0_5

21. Polzer B, Medoro G, Pasch S, Fontana F, Zorzino L, Pestka A, Andergassen U, Meier-Stiegen F, Czyz ZT, Alberter B, Treitschke S, Schamberger T, Sergio M, Bregola G, Doffini A, Gianni S, Calanca A, Signorini G, Bolognesi C, Hartmann A, Fasching PA, Sandri MT, Rack B, Fehm T, Giorgini G, Manaresi N, Klein CA (2014) Molecular profiling of single circulating tumor cells with diagnostic intention. EMBO Mol Med 6(11): 1371–1386. https://doi.org/10.15252/emmm.201404033

Chapter 7

Isolation and Genomic Analysis of Circulating Tumor Cell Clusters in Cancer Patients

Carolina Reduzzi, Marta Vismara, Thomas Schamberger, Marco Silvestri, Rosita Motta, Bernhard M. Polzer, and Vera Cappelletti

Abstract

The role of circulating tumor cell (CTC) clusters in the metastatic dissemination process is gaining increased attention. Besides homotypic clusters, heterotypic clusters that contain tumor cells admixed with normal cells are frequently observed in patients with solid tumors. Current methods used for cluster detection and enumeration do not allow an accurate estimation of the relative fractions of tumor cells. Here we describe a method for estimating tumor fraction of clusters including isolation and collection of single clusters, assessment of copy number alterations of single clusters by low-pass whole genome sequencing, and bioinformatic analysis of sequencing data.

Key words CTC clusters, Micromanipulation, Genomic analysis, Copy number alteration, Tumor fraction prediction, Low-pass whole genome sequencing

1 Introduction

The role of circulating tumor cell (CTC) clusters in the metastatic dissemination process is gaining increased attention [1]. CTC clusters, either representing a grouping of CTCs, i.e., homotypic clusters, or associations between CTCs and accessory normal cells, i.e., heterotypic clusters, do have distinct features with respect to single CTCs and possibly a different biological role and clinical relevance in the metastatic dissemination [2–4]. As such, CTC clusters represent a unique biomarker which deserves to be separately studied.

No standardized assay is currently available for CTC-cluster enumeration and literature data addressing the clinical role of CTC clusters have been produced with an array of different approaches including CellSearch® [5], the only FDA-approved method for enumeration of single CTCs.

Moreover, due to their heterotypic nature, CTC clusters pose additional challenges beyond simple isolation and enumeration, dealing with specifying their composition. Here we describe a

Miodrag Guzvić (ed.), *Single Cell Analysis: Methods and Protocols*, Methods in Molecular Biology, vol. 2752,
https://doi.org/10.1007/978-1-0716-3621-3_7,

comprehensive protocol spanning from CTC-cluster collection to prediction of their tumor fraction, which defines the percent content of tumor cells in each single cluster based on the content of genomically aberrant cells estimated by copy number alteration (CNA) profiles generated by shallow whole genome sequencing.

The protocol has been developed for breast cancer patients [6] following the observation of a higher number of CTC cluster in women with early breast cancer than in those with metastasis [7], as a first step toward clarifying this biologically and clinically intriguing observation. Although not yet validated in other tumor types, it can be simply extended by running a new calibration curve with ad hoc cell lines.

2 Materials

2.1 CTC-Cluster Enrichment and Collection of Single CTC Clusters (See Subheading 3.1)

- Blood collection tubes: K2-EDTA BD Vacutainer tubes (BD).
- Blood filtration devices: ScreenCell CYTO kit (ScreenCell).
- 15 mL and 50 mL sterile conical tubes.
- Phosphate-buffered saline (PBS) 1×.
- Parafilm.
- May Grunwald (Merck Millipore).
- Giemsa (Merck Millipore).
- Tap water.
- Paper towels.
- Tweezer.
- Serological pipettes (5 and 10 mL) with pipette controller.
- 200 μL micropipette tips with micropipette.
- 4 °C refrigerator.
- Inverted microscope (e.g., Olympus IX81).
- Patchman NP2 micromanipulator (Eppendorf).
- DMZ Capillary puller (Zeitz-Instruments Vertriebs GmbH).
- CellTram Air pump (Eppendorf).
- Glass capillaries: micro-hematocrit capillary, non-heparinized length 75 mm × 1.1/1.2 mm (Nunc).
- BSA (Sigma-Aldrich).
- Fetal bovine serum (FBS), sera plus, (PAN Biotech).
- Nunc LabTek Chamber Slides; 8 fields (Thermo Fisher Scientific).
- Ice.
- 200 μL MAXYMum Recovery® PCR Tubes (Axygen).

2.2 Genomic Analysis of CTC Clusters: WGA, QC, and Low-Pass CNA (See Subheading 3.2)

2.2.1 Whole Genome Amplification (WGA)

- Ampli1™ WGA Kit (Menarini Silicon Biosystems; *see* **Note 1**).
- *DNA-OFF DNA removal agent (MP Biomedicals).*
- Thermal Cycler (Applied Biosystems GeneAmp 9700 or superior; *see* **Note 2**).
- Set of dedicated micropipettes (Gilson).
- PCR microcentrifuge tube 0.2 mL (Amplitube™ thin wall reaction tube 200 μL flat snap cap, Starlab).
- Dual-filter/barrier tips (Eppendorf).
- Mini Centrifuge suitable for 0.2 mL PCR tubes.
- Laminar flow hood.
- −20 °C storage freezer.
- Vortex mixer.
- *See* **Notes 3** and **4**.

2.2.2 Quality Control (QC)

- Ampli1™ QC Kit (Menarini Silicon Biosystems).
- Dedicated micropipette set for post-PCR reactions.
- 0.2 mL PCR tubes.
- 1.5 mL PCR-clean tubes.
- Microcentrifuge for 1.5 and 0.2 mL tubes.
- Barrier tips.
- −20 °C Storage Freezer
- Thermal Cycler.
- Agilent 2100 Bioanalyzer (Agilent Technologies).
- Agilent DNA 1000 Kit (Agilent Technologies).

2.2.3 LowPass Copy Number Alteration (LpCNA) Sequencing

- Ampli1™ LowPass Kit for Ion Torrent (Menarini Silicon Biosystems; *see* **Note 6**).
- *DNA OFF DNA removal agent (MP Biomedicals).*
- Thermal Cycler (Applied Biosystems GeneAmp 9700 or superior; *see* **Note 2**).
- Ethanol Bio Ultra, for molecular biology, ≥99.8%.
- Nuclease-free water (Molecular Biology Grade).
- Low TE: 10 mM Tris–HCl, pH 8.0, 0.1 mM EDTA).
- 0.2 mL PCR tubes (Axygen Maxymum Recovery thin wall PCR tubes with flat cap)
- Aerosol-resistant tips and pipettes range from 1 to 1000 μL.
- SPRI select reagent (Beckman Coulter).

- Magnetic rack for 0.2 mL tubes (DynaMag-96 Side magnet, Thermo Fisher Scientific).
- Programmable thermal cycler with adjustable ramp rate, operating within manufacturer's specifications.
- Dedicated micropipette sets for pre-PCR and post-PCR reactions.
- Filter tips (Gilson Diamond sterilized).
- Mini centrifuge for 0.2 mL tubes.
- Vortex.
- Agilent 2100 Bioanalyzer (Agilent Technologies).
- Agilent High Sensitivity DNA Kit (Agilent Technologies).
- E-Gel Precast Agarose Electrophoresis Systems (Thermo Fisher Scientific).
- E-Gel Size Select Agarose Gels 2%, (Thermo Fisher Scientific).
- 50 pb DNA Ladder.
- Laminar flow hood.
- Tape station (Agilent Technologies).

2.3 Bioinformatic Analysis and Interpretation of Sequencing Data

2.3.1 Pre-processing and Alignment

- Torrent Suite™.

2.3.2 Alignment Quality Control

- Qualimap [8].
- FastQC (http://www.bioinformatics.babraham.ac.uk/projects/fastqc).

2.3.3 Copy Number Alteration Analysis and Tumor Fraction Estimation

- R.
  - HMMcopy package.
  - GenomeInfoDb package.
  - GenomicRanges package.
- ichorCNA [9].

2.3.4 Data Visualization and Operative System

- Bash.
- R.
- Linux/Unix.

3 Methods

3.1 CTC-Cluster Enrichment and Picking of Single CTC Clusters

1. Collect 9 mL of peripheral venous whole blood in K2-EDTA tubes (*see* **Notes 7** and **8**). Invert the tubes 10 times and store the blood at 4 °C until blood filtration, for a maximum of 4 h (*see* **Note 9**).
2. Open one ScreenCell filtration unit (included in the ScreenCell CYTO kit) and place it on the bench top with the adhesive protective membrane facing down (*see* **Note 10**).
3. Prepare the dilution buffer included in the ScreenCell CYTO kit (FC2 buffer) by fully screwing the cap on the vial until the clap of the yellow plunger in the cap is opened (the molded tamper-evident ring will break). Shake the vial by hand at least 10 times and check the buffer pH using the pH indicator strip provided in the kit. If needed, add up to 3 μL of 30% NaOH (included in the kit) to reach a pH between 6.7 and 7.3. Use the dilution buffer within 24 h.
4. Transfer 3 mL of blood in a 15 mL sterile conical tube, add 4 mL of the prepared FC2 dilution buffer (*see* **step 3**), gently invert the tube 10 times, and incubate at room temperature for 8 min.
5. At the end of the incubation, transfer the 7 mL of diluted blood in the module A of the filtration unit prepared in **step 2**. Remove the adhesive protecting membrane from the lower part of the filtration unit (module B), being careful not to spill the blood contained in module A. Insert module B onto an empty blood collection tube provided in the ScreenCell CYTO kit (module C) and push down completely to start the filtration. When the blood reaches the yellow line of module A, add 1.6 mL of PBS 1× into module A. Wait until all the liquid has gone through the membrane, it should take approximately 3 min (*see* **Note 11**).
6. Open the device by twisting module A and eject the isolation support (IS, formed by the filtration membrane and an inox O-ring) onto a piece of absorbent paper. Wash the IS by putting 70 μL of PBS on the IS and let the PBS be absorbed by the paper, through the IS. Using the tweezer, put the filter on a piece of well-extended parafilm and let the IS 1 h at 37 °C or dry overnight at room temperature. Then proceed with staining.
7. Incubate the filter in May Grunwald for 2.5 min at room temperature, transfer the filter into diluted May Grunwald (diluted 1:2 with distilled water pH 7.0), and incubate for 2.5 min. Rinse the IS in tap water 2–3 times. Incubate for

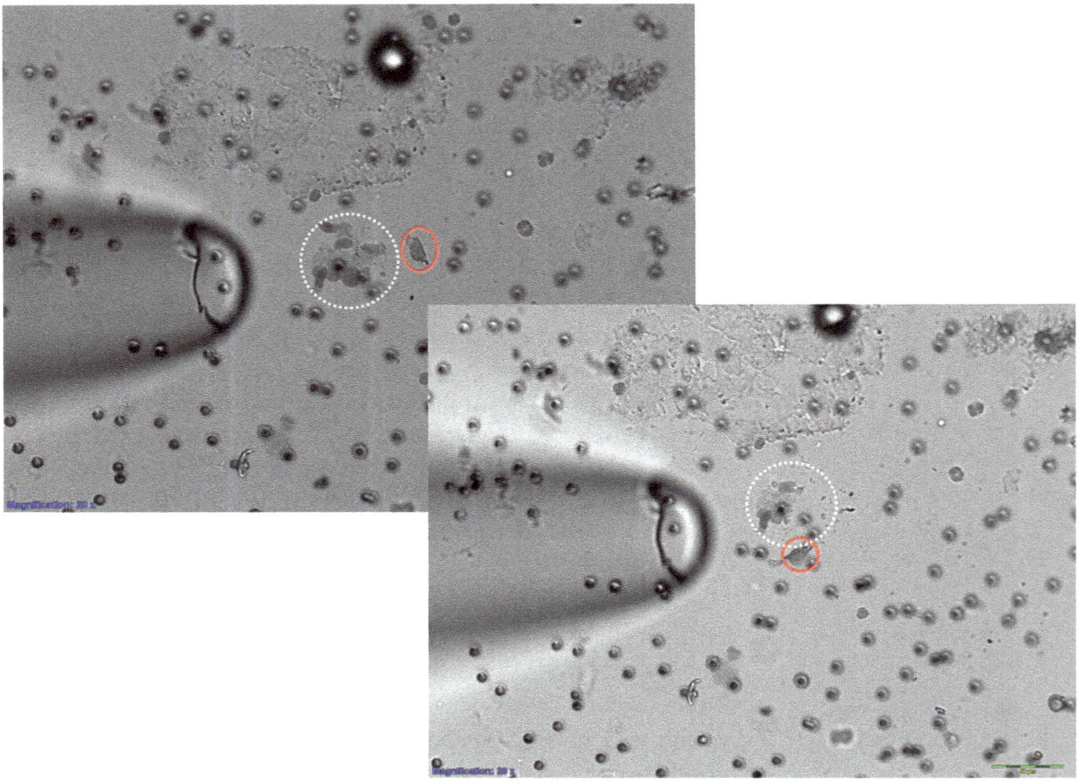

Fig. 1 The pictures show a cluster (dotted white circle) before and after cell detachment with the micromanipulator. The red circles indicate two single cells detached from the cluster

10 min the IS in Giemsa diluted 1:10 with distilled water pH 7.0, then rinse in tap water (*see* **Note 12**).

8. Let the IS air-dry for 15 min. Store the IS at 4 °C for 1–2 weeks maximum and proceed with the procedure for CTC clusters isolation as follows.
9. Prepare 0.2 mL PCR tubes containing 2 μL of lysis buffer from the WGA kit (see Subheading 3.2) and store them on ice.
10. Put a glass capillary to the Zeitz capillary puller. Two pick capillaries are formed from one glass capillary, each with an opening of 30–50 μm (*see* **Note 13**).
11. Put the IS onto the microscope and add approximately 0.5 mL of 1×PBS to the slide.
12. Coat one field of an 8-field chamber slide with BSA, add 200 μL 1×PBS (= picking field), and place the slide next to the IS on the microscope.
13. Add by hand FBS into the picking capillary and connect the capillary to the CellTram Air pump (*see* **Note 14**).

14. Focus the picking field and move the picking capillary by using the Patch Man joystick into the middle of the visible field. Lower the capillary to the surface of the picking field and flush out the FBS (not all) using the air pump. Lift the capillary out of the picking field (*see* **Note 15**).
15. Take 1 μL of PBS from the picking field and transfer into a 0.2 mL PCR tube containing 2 μL lysis buffer from the WGA kit (*see* Subheading 3.2) and store it on ice until the end of the isolation procedure. This refers to picking (negative) control.
16. Focus a target CTC cluster, lower the picking capillary in front of the cluster. Try to remove the cluster by pushing or scratching it with the glass capillary. Be careful and do not destroy the cells/cluster (Fig. 1) (*see* **Note 16**).
17. Aspirate the target cluster into the glass capillary, lift the capillary out of the sample slide, focus the picking field, and lower the glass capillary to the surface (*see* **Note 17**).
18. Take a 10 μL pipette including 10 μL pipette tip. Flush the picked cluster out of the capillary and collect it with the pipette within 1 μL (*see* **Note 18**).
19. Transfer the isolated cluster into a 0.2 mL PCR tube containing 2 μL of the lysis buffer from the WGA kit prepared in **step 9** and store on ice until the end of the isolation procedure, and then continue with the WGA procedure (*see* **Note 19**).

3.2 Genomic Analysis of CTC Clusters

3.2.1 Whole Genome Amplification (WGA)

The Ampli1™ WGA method was originally designed to amplify DNA from a single cell, but is also suitable for amplification of CTC clusters whose cell number ranges from 2 cells to 1000. Due to the extreme sensitivity of the WGA, it is important to avoid contamination by exogenous DNA during the CTC-picking step.

The Ampli1™ WGA method is based on a ligation-mediated adaptor linker PCR technique, which assures a balanced amplification due to restriction enzyme-promoted fragmentation of the genome. The workflow for Ampli1™ WGA includes five steps that may be executed in 1 day.

All reagents necessary to successfully execute the WGA procedure are provided in the Ampli1™ WGA Kit.

1. Prepare control samples.

   It is strictly recommended to process the following controls along with samples in each run of Ampli1™ WGA procedure:

   - Two no-cell controls: by adding 1 μL of Ampli1™ Water.
   - One positive control: by adding 1 μL of purified DNA (1 ng/μL) extracted from a cell line. For this step use a dedicated and decontaminated 10 μL micropipette. It is

strongly recommended to process the positive control as the last sample of each step.

2. Open the Ampl1™WGA kit in the dedicated laboratory space and proceed with the four steps according to kit instructions (*see* **Notes 3–6**).
3. Proceed with STEP 1 (cell lysis) using the PCR tubes deriving from the picking procedure, containing CTC clusters and lysis buffer; prepare new PCR tubes for the negative and positive WGA controls. Continue with STEP 2A (DNA digestion) and STEP 2B always referring to the Ampli1™WGA kit user manual v6 (*see* **Note 20** for STEP 2B).
4. Continue with STEP 3 (ligation) and STEP 4 (primary PCR) as described in the Ampli1™WGA kit user manual v6 (*see* **Note 21** for the final step).

3.2.2 Quality Control (QC) of WGA Products

The Ampli1™QC kit contains reagents for a PCR-based assay useful for the qualitative evaluation of the obtained Ampli1™WGA products (for human-origin samples). It predicts the suitability of each sample for downstream analyses and is based on a multiplex PCR for four markers. Positivity of at least two markers predicts successful genome-wide analyses to detect copy number alterations (CNA).

Store the Ampli1™ QC Kit at −15/−20 °C. Transfer the tube containing the enzyme to ice just prior to use. Other kit components should be thawed, stored on ice, and briefly vortexed before use (*see* **Note 22**).

1. It is recommended to prepare the following controls for Ampli1™ QC reaction:
   - Blank control: by adding 1 μL of PCR water.
   - Positive control: by adding 1 μL of 100 ng/μL human genomic DNA.
2. Prepare the reaction mixture according to the instructions of the Ampli1™QC kit. Add 1 μL of Ampli1™WGA product, prepare a blank and a positive control, and incubate in the thermal cycler as indicated.
3. For the qualitative evaluation of PCR products, we load 1 μL of the final product on the Agilent DNA1000 chip on the Bioanalyzer Agilent 2100.

3.2.3 Low Pass CNA Sequencing

Ampli1™ LowPass Kit is designed to generate multiplexed, sequencing-ready libraries in a streamlined single-day protocol, to detect chromosomal aneuploidies and CNA, by low-pass whole genome sequencing using the Ion PGM or the Ion S5 systems (*see* **Note 6**). Libraries are created through a single-reaction step and offer the possibility of generating up to 48 barcoded libraries suitable for both Ion Torrent NGS platforms.

The kit offers a robust and reproducible method for shallow whole genome to assess genome-wide copy number alterations.

Upon receipt, store the Ampli1™ LowPass Kit at −20 °C. Transfer the tube containing the enzyme to ice just prior to use, add the enzyme to the reaction mix as the last component, and immediately put the tube containing the enzyme back at −20 °C. The other kit components should be thawed, stored on ice, and briefly vortexed before use.

Open the Ampl1™ Low pass Kit in the dedicated laboratory space (*see* **Notes 23** and **24**) and proceed through four steps according to kit instructions.

1. Proceed with STEP 1 (Clean up of Ampli1™ WGA product) referring to the user manual of Ampli1™ LowPass Kit v5 (*see* **Note 25**).
2. Continue with STEP 2 (Barcoding Reaction) according to the user manual of Ampli1™ LowPass Kit v5 (*see* **Note 26**).
3. Thereafter, proceed with STEP 3 (Clean up of Final Library) according to the user manual of Ampli1™ LowPass Kit v5 (*see* **Note 27**).
4. Next, perform STEP 4 (Pool Barcode Libraries by Tape Station) as specified in the user manual v5 (*see* **Note 28**).
5. Finally go to STEP 5 (Size Selection) referring to the user manual of Ampli1™ LowPass Kit v5 and to **Note 29** and conclude with STEP 6 (Clean up of Final Library) according to the user manual of Ampli1™ LowPass Kit v5 and to **Note 30**.

Refer to your sequencing service provider for the final steps:

- STEP 7: Final Library Quantification.
- STEP 8: Sequencing setup.

3.3 Bioinformatic Analysis and Interpretation of Sequencing Data

3.3.1 Pre-processing and Alignment (See Note 31)

1. Open Torrent Suite software and perform the pre-processing steps on raw sequences as indicated by the manufacturer:
   - Quality check.
   - Alignment to hg19 (TMAP aligner).
2. Obtain alignment file (.bam) from Torrent Suite. Exclude samples not passed quality control and/or samples with aligned reads counts lower than 400,000.

3.3.2 Alignment Quality Control

1. Perform alignment quality control using FastqQC with standard settings. Do not consider “Per base sequence content,” “Per sequence GC content,” “Sequence length distribution,”

and “Overrepresented sequences” parameters during the analysis. Exclude samples with >1 “Fail” returned by software.

2. Perform alignment quality control using Qualimap selecting “BAM QC” as analysis type. Exclude samples with “mapping quality mean” < 50.

3.3.3 Copy Number Alteration Analysis and Tumor Fraction Estimation

1. Run ichorCNA on alignment file (.bam) with the following settings (*see* **Note 32**):
   - Window = 1 Mb.
   - Ploidy = 2.
   - estimatePloidy = TRUE.
   - estimateNormal = TRUE.
   - normalPanel = TRUE.
   - normal = c(0.1, 0.2, 0.3, 0.4, 0.5, 0.6, 0.7, 0.8, 0.9).
2. Retrieve the Copy number alteration calls for each segmented genomic region (1 Mb) and estimated tumor fraction (TF) from <sampleID>.seg.txt and < sampleID>.params.txt files, respectively (*see* **Note 33**).
3. Do not consider chr19 during the evaluation of CNA profile (*see* **Note 34**). Classify as unclear the CNA profiles showing one of the following features:
   - Normal profile but only 1 genomic region with amplification/deletion lower than 125 Mb.
   - Normal profile but sum of amplification/deletion of different genomic regions lower than 375 Mb.

   Classify all CNA profiles with alterations above these thresholds as aberrant.
4. Classify samples as Normal, Tumor, or Unclear based on criteria shown in the following table (*see* **Notes 35** and **36**). Please see Fig. 2 for an example of two clusters containing cells with aberrant DANA and one with normal diploid cells (Table 1).

4 Notes

1. Store all reagents and enzymes according to the guidelines specified in the corresponding data sheets provided by the manufacturer, in a separate box from the DNA samples. Avoid using reagents that have expired or reagents for which you suspect an improper storage. Completely thaw and pre-chill to 4 °C all reagents before preparation of reaction master mix, as use of non-properly chilled reagents may cause errors in pipette volumes. Since multiple freezing/thawing

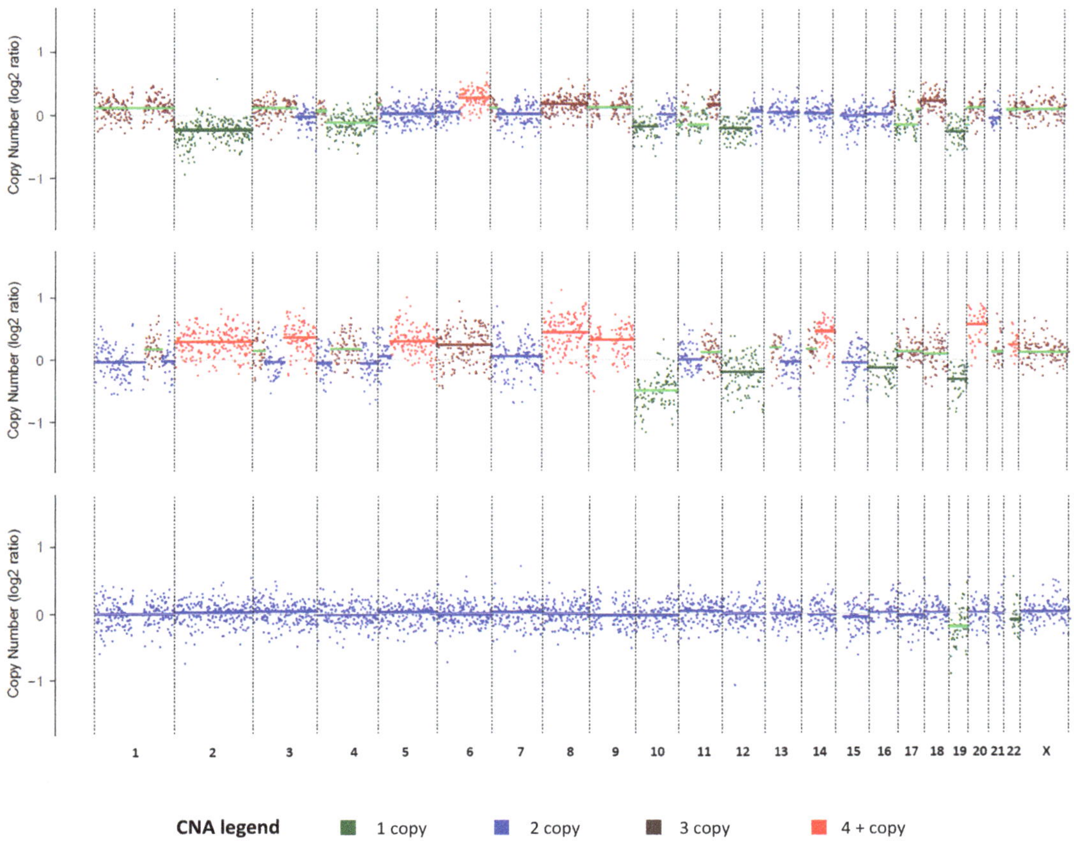

Fig. 2 CNA profiles obtained with low-pass WGS from 3 CTC clusters: two with aberrant CNA profiles (upper) and one with normal CNA profile (bottom). Color codes refer to 1 copy, 2 copies, 3 copies, and more than 4 copies for each single genomic region

Table 1
CNA classification of isolated clusters

TF	+	CNA profile	=	Final output
$0 \leq TF \leq 0.05$	+	Normal/aberrant/unclear	=	Normal CTC cluster
$0.05 < TF \leq 1$	+	Aberrant	=	Aberrant CTC cluster
$0.05 < TF \leq 1$	+	Unclear	=	Unclear CTC cluster

cycles of temperature-sensitive reagents (i.e. enzymes, ATP, and dNTPs) may impair their performance, we recommend generating smaller aliquots to prevent their decomposition. Put special attention to the water used for dilutions and controls by using DNAse-free sterile water (provided in the kit) stored in small aliquots and use a new aliquot for each run.

2. Use a programmable thermocycler with adjustable ramping rates.
3. Use only dual-filter/barrier pipette tips to prevent the cross-contamination of the samples. We recommend using flat-lid PCR reaction tubes with shield cap which facilitates labeling of samples.
4. It is recommended to designate a separate laboratory or working space dedicated to the WGA only. Always pay great attention to maintain high standards of purity and sterility in the working space, which must absolutely be physically separated from areas of the laboratory used for other molecular biology activities, in particular post-PCR processing of samples, as this could result in contaminations. Ideally, all steps of the procedure should be performed under a dedicated laminar flow hood equipped with the full set of laboratory equipment (i.e., dual filter tips, dedicated set of pipettes, which must be used exclusively used for the Ampi1™ WGA) and consumables. Pipettes are particularly critical for a successful reaction and must be calibrated and properly handled. Please refer to manufacturer instructions for pipette-calibration and for pipette decontamination. We recommend decontamination steps to be performed each time experiments are run after a long interruption time. Before starting each experiment always clean the top of the laminar flow hood with the DNA-off solution followed by sterile water and 70% ethanol. Always use disposable lab coats and nitrile/powder-free gloves. To save time all incubation steps of the protocol should be pre-programmed on the thermal cycler.
5. It is critical that the operators are experienced in pipetting very small volumes. Several tips may be useful: always pipette the small volumes of reaction mixes gently, spin the tubes using a microcentrifuge, pipette the desired volume of the reaction mix onto the wall of the tube, without disturbing the fluid inside, and collect it on the bottom of the tube with a short centrifugation step. Samples should be briefly centrifuged after each subsequent step of the procedure. To compensate for volume losses during pipetting, it is suggested to include a 5% volume excess in each reaction mix.
6. You can also use the Ampli1™ LowPass Kit for Illumina (Menarini Silicon Biosystems).
7. Each filtration unit can process up to 3 mL of blood. To process a higher volume, collect the blood in multiples of 3 and process it using multiple filtration kits.
8. To avoid the risk of epithelial skin cell contamination, discard the first ml of blood or collect the tubes for CTC analysis after collection of blood samples for other analyses.

9. Filter the blood as soon as possible, better within 1 h. The longer the blood is kept before filtration, the higher is the possibility that clots will form, causing the stop of the filtration before the end of the sample.
10. 50 mL tube racks can be used to hold the filtration units in place during processing.
11. In case blood filtration stops, it is possible to try to restart it by inserting the filtration unit into a new empty blood collection tube (module C). No additional tubes are provided with the kit, but it is possible to use conventional 10 mL EDTA tubes.
12. In alternative to May Grunwald Giemsa stain, it is possible to stain with hematoxylin and eosin using the following protocol. Put 70 μL of filtered hematoxylin on the IS positioned on the parafilm. Incubate for 1 min and then, using the tweezers, move the IS on the absorbent paper to absorb the hematoxylin. Immediately rinse the IS by dipping it in a 50 mL conical tube containing tap water, for 2/3 times, using the tweezers to hold the IS. Put the filter on absorbent paper to absorb the excess of water and put it back on the parafilm. Put 70 μL of eosin on the IS and incubate for 30 s. Put the IS on the absorbent paper and absorb the eosin. Immediately rinse the IS by dipping it in a 50 mL conical tube containing tap water, for 2/3 times, using the tweezers to hold the IS. Put the filter on absorbent paper to absorb the excess of water.
13. You can use one capillary per patient slide. Be careful not to destroy the tip; use gloves.
14. Fill 50 μL of FBS in a 200 μL tube. Take the capillary and put it into the FBS. FBS absorbs itself. Do not touch the tube surface with the tip.
15. Do not flush out too much of the FBS. A little amount of FBS should stay to prevent the cell attachment to the inner surface of the picking capillary. If you have too little FBS in the capillary, you will not be able to aspirate a cell. If you have too much FBS in, the capillary pressure is too high and you might lose the isolated cell.
16. If there are some other unwanted cells near the target cell, remove the cells using the glass capillary by scratching, but be sure that no one of the unwanted cell is attached in or outside the glass capillary.
17. Do not aspirate the cell too less or too much into the picking capillary. Too less: cell lost by moving out; too much: high pressure. Should you aspirate too much, put out the liquid very carefully until you can see the cluster.
18. To ensure a secure and stable hold, it is helpful to lean your forearm against the microscope. If necessary, a box can also be

placed on the table. Then place your elbow on the box and lean your forearm against the microscope. Caution: do not touch any moving parts such as the microscope table with your forearm.

19. This is a possible stopping point. Cells can be stored at −20 °C until WGA.
20. This step can be done in parallel with STEP 1/STEP 2A by using a different thermal cycler. Otherwise, the two steps must be done subsequently in the same thermal cycler storing the samples at +2–8 °C. This step is a pre-preparation of a component for the reaction mix of STEP 3.
21. Once the program is ended, remove the tunes and store them at −20 °C in a freezer located in a separate lab space dedicated to downstream analysis of the amplified products.
22. Ampli1™ QC Kit is a POST WGA amplification kit. The reaction must be prepared in the PCR working area, and NOT in WGA working area. Pipettes for WGA should not be used for this step. To save time all incubation steps of the protocol should be pre-programmed on the thermal cycler.
23. In order to prevent any contamination, it is strongly recommended to wear powder-free nitrile gloves for all protocol steps, use only freshly opened plastic ware, dedicate a separate working area, and use separate set pipettes for pre-PCR and for library preparations. We recommend using a laminar-flow hood and cleaning the top with DNA-off, sterile water, and ethanol 70° before starting. To save time, all incubation steps of the protocol should be pre-programmed on the thermal cycler. Before starting, clean the laminar flow hood, prepare the magnet, prepare the requested volume of ethanol 80%, the dedicated pipettes, the LowTE buffer, the labeled 0.2 mL tubes, and do not forget to vortex the beads making sure to obtain a homogeneous solution. We suggest processing 12 samples in each single experiment.
24. To compensate for volume losses during pipetting, it is suggested to include a 5% volume excess in each reaction mix. When handling the barcodes, open one tube at time to prevent any cross-contamination. Touch the side of tube with the pipette tip and slowly dispense reagent on the side to form a droplet.
25. Be very careful during removal of the supernatant to avoid touching the beads loaded with the Ampli1™ WGA fragments. Before discarding the supernatant make sure that there are no beads in the tip. Should the pipette tip contain beads, carefully discharge the supernatant back into its tube and wait a few minutes to allow the optimal capture of the beads by the magnet before removing the supernatant. For the first washing

step, 200 μL of ethanol 80% is added to beads; before proceeding with the second washing step, remove 150 μL and replace them with 150 μL of fresh ethanol, thereafter remove the entire supernatant, by removing twice 110 μL of ethanol, and remove all residual ethanol using a 10 μL micropipette. Do not leave the beads get too much dry, to avoid reducing the efficiency of the TE-elution step.

26. Always check that the PCR program is still stored in the Thermal Cycler memory. Prepare the mix for 12.5 samples. Thaw all the reagents except the Taq enzyme. Remove the Taq enzyme from the freezer only last minute. Thaw, vortex, and briefly spin the tubes containing the barcodes.

27. Use the same precautions as for step 1! If you decide not to proceed immediately with step 4b, store the samples at 4 °C (for 24 h storage) or at −20 °C (for long-time storage).

28. After the 1:5 dilution of the samples with LowTE, briefly vortex and spin. Annotate the concentration (ng/μl) by selecting the area between 300 and 450 bp for each sample profile on the TapeStation. Prepare a final equimolar pool of barcoded libraries based on the concentrations measured by the TapeStation. For the pool preparation step, order the samples based on their concentrations starting from the lowest one and divide them into two sets. Use each half of the samples to prepare pools paying attention to avoid having too large volume differences between the samples. This can be achieved by slightly adjusting the volumes to maintain the equimolarity, but avoid pipetting too small volumes. Do not pipette less than 1 μL per sample to avoid inaccurate volumes. The final pool should be at least 15 ng/μL and no more than 38 ng/μL in a final volume of 44 μL. Each pool contains a maximum of 24 samples.

29. Where specified in the manufacturer's instructions, use LowTE instead of water (nuclease-free water) and before loading the samples add to the wells 20 μL instead of 25 μL. The run should be stopped when the 350 bp band starts to exit from the central well (loaded with the marker) and the 450 bp band starts entering the well. For this step, refer to the fluorescent bands 350 and 450 from the reference lane, and monitor very well the run-in order to avoid missing the best time for stopping the run. Now you can collect the pool fraction containing fragments ranging from 350 to 450 bp. Add a small volume of LowTE to the wells and recover the pool fraction touching the bottom of the well to be sure to collect the entire volume; do not worry if you end up breaking the gel. The recovered volume should be around 50–60 μL. Spin the pool at 14,000 × *g* for 10 min to remove gel leftovers and collect the

supernatant leaving a small volume of liquid on the bottom of the tube.

30. Use the same precaution as in step 1! Store the samples at 4 °C or at −20 °C (for long-time storage) if you decide not to proceed with the next step.
31. In the case of samples processed using Illumina NGS platform, consider the following Pre-processing & alignment step.
    - Perform raw sequence quality control using FastqQC with standard settings. Do not consider "Per base sequence content," "Per sequence GC content," "Sequence length distribution," and "Overrepresented sequences" parameters during the analysis. Exclude samples with >1 "Fail" returned by software.
    - Perform raw sequencing trimming and adapter clipping using trimmomatic.
    - Download reference genome hg19 from https://hgdownload.soe.ucsc.edu/downloads.html.
    - Perform alignment to human reference genome hg19 using BWA-MEM with standard settings.
32. ichorCNA gives the possibility to create reproducible and scalable pipeline using snakemake (see https://github.com/broadinstitute/ichorCNA/wiki/SnakeMake-pipeline-for-ichorCNA).
33. The <sampleID>.seg.txt file reports copy number state in terms of median log2 ratio and absolute copy number. In order to classify a CNA profile as normal, aberrant, or unclear (see **step 3**), only median log2 ratio could be considered using the following threshold:
    - Median log2 ratio ≥ 0.1: Gain.
    - Median log2 ratio ≤ −0.1: Loss.
    - −0.1 < Median log2 ratio < 0.1: Normal.

    Complete CNA profile image is reported in <sampleID>_genomeWide.pdf file.
34. Chr19 should not be considered during the evaluation of CNA profile due to its biased deletion calls associated with the high CG base percentage [10].
35. Use R software (or other programming language for data analysis) to handle and process all data from **steps 2–4**.
36. It is important to note that the TF value returned by ichorCNA is estimated only depending on the "amount" of genomic aberration not corrected by the different degrees of CNAs characterizing cancer cells within the CTC clusters. To cope with this problem, we have previously published a method to

obtain a TF value corrected for the different degrees of DNA alteration and for sequencing coverage [6].

References

1. Castro-Giner F, Aceto N (2020) Tracking cancer progression: from circulating tumor cells to metastasis. Genome Med 12(1):31. https://doi.org/10.1186/s13073-020-00728-3
2. Aceto N (2020) Bring along your friends: homotypic and heterotypic circulating tumor cell clustering to accelerate metastasis. Biom J 43(1):18–23. https://doi.org/10.1016/j.bj.2019.11.002
3. Heeke S, Mograbi B, Alix-Panabieres C, Hofman P (2019) Never travel alone: the crosstalk of circulating tumor cells and the blood microenvironment. Cell 8(7). https://doi.org/10.3390/cells8070714
4. Schuster E, Taftaf R, Reduzzi C, Albert MK, Romero-Calvo I, Liu H (2021) Better together: circulating tumor cell clustering in metastatic cancer. Trends Cancer 7(11): 1020–1032. https://doi.org/10.1016/j.trecan.2021.07.001
5. Pineiro R, Martinez-Pena I, Lopez-Lopez R (2020) Relevance of CTC clusters in breast cancer metastasis. Adv Exp Med Biol 1220: 93–115. https://doi.org/10.1007/978-3-030-35805-1_7
6. Silvestri M, Reduzzi C, Feliciello G, Vismara M, Schamberger T, Kostler C, Motta R, Calza S, Ferraris C, Vingiani A, Pruneri G, Daidone MG, Klein CA, Polzer B, Cappelletti V (2021) Detection of Genomically aberrant cells within circulating tumor microemboli (CTMs) isolated from early-stage breast cancer patients. Cancers (Basel) 13(6). https://doi.org/10.3390/cancers13061409
7. Reduzzi C, Di Cosimo S, Gerratana L, Motta R, Martinetti A, Vingiani A, D'Amico P, Zhang Y, Vismara M, Depretto C, Scaperrotta G, Folli S, Pruneri G, Cristofanilli M, Daidone MG, Cappelletti V (2021) Circulating tumor cell clusters are frequently detected in women with early-stage breast cancer. Cancers (Basel) 13(10). https://doi.org/10.3390/cancers13102356
8. Okonechnikov K, Conesa A, Garcia-Alcalde F (2016) Qualimap 2: advanced multi-sample quality control for high-throughput sequencing data. Bioinformatics 32(2):292–294. https://doi.org/10.1093/bioinformatics/btv566
9. Adalsteinsson VA, Ha G, Freeman SS, Choudhury AD, Stover DG, Parsons HA, Gydush G, Reed SC, Rotem D, Rhoades J, Loginov D, Livitz D, Rosebrock D, Leshchiner I, Kim J, Stewart C, Rosenberg M, Francis JM, Zhang CZ, Cohen O, Oh C, Ding H, Polak P, Lloyd M, Mahmud S, Helvie K, Merrill MS, Santiago RA, O'Connor EP, Jeong SH, Leeson R, Barry RM, Kramkowski JF, Zhang Z, Polacek L, Lohr JG, Schleicher M, Lipscomb E, Saltzman A, Oliver NM, Marini L, Waks AG, Harshman LC, Tolaney SM, Van Allen EM, Winer EP, Lin NU, Nakabayashi M, Taplin ME, Johannessen CM, Garraway LA, Golub TR, Boehm JS, Wagle N, Getz G, Love JC, Meyerson M (2017) Scalable whole-exome sequencing of cell-free DNA reveals high concordance with metastatic tumors. Nat Commun 8(1):1324. https://doi.org/10.1038/s41467-017-00965-y
10. Zhao M, Wang Q, Wang Q, Jia P, Zhao Z (2013) Computational tools for copy number variation (CNV) detection using next-generation sequencing data: features and perspectives. BMC Bioinform 14(Suppl 11). https://doi.org/10.1186/1471-2105-14-S11-S1

Chapter 8

Establishing Single-Cell Clones from In Vitro-Cultured Circulating Tumor Cells

Teng Teng and Min Yu

Abstract

Cancer is a common health problem with more than 90% of deaths due to metastases. Circulating tumor cells (CTCs) contain precursors that can initiate metastases. However, CTCs are rare, heterogeneous, and difficult to expand in culture. We have previously created CTC-derived cell lines from stage IV breast cancer patients. These CTC lines were used to establish single-cell CTC clones using flow cytometry cell sorting.

Key words Circulating tumor cells, Single-cell clones, CTC line culture, Breast cancer, Heterogeneity

1 Introduction

Cancer is a common disease worldwide, with survival rates varying based on the type of cancer and geographical region [1]. For those who don't survive, metastatic tumors at sites distant from the primary tumors are the main cause of cancer-related death [2]. The "seeds" initiating the multi-step cascade of hematogenous metastasis are a population of tumor cells in the circulatory system, called circulating tumor cells (CTCs). CTCs originate from primary or metastatic lesions, encounter a new set of stresses in the circulation during their transit to distant sites, and extravasate into the secondary organs. Here, some of these CTCs eventually create a new metastatic lesion, or remain dormant [3, 4]. To demonstrate this tumor-initiating capacity of CTCs in vivo, scientists recently established CDX (CTC-derived xenograft) mouse models [5]. Many studies have demonstrated the potential value of CTCs in diagnosing disease, monitoring progression, predicting patient outcomes, and designing therapeutic targets [6]. Recently, clinical trials implemented CTC enumeration in order to monitor the chemotherapeutic responses of prostate cancer patients [7]. Therefore, CTCs hold promise as a "liquid biopsy" for cancer patients.

Miodrag Gužvić (ed.), *Single Cell Analysis: Methods and Protocols*, Methods in Molecular Biology, vol. 2752,
https://doi.org/10.1007/978-1-0716-3621-3_8, © Springer Science+Business Media, LLC, part of Springer Nature 2024

Traditionally, biopsies require surgery, which is invasive. In contrast, clinicians can isolate CTCs from the peripheral blood via venipuncture, which is much less invasive. However, CTCs are extremely rare and admixed in a huge number of normal blood cells, presenting a significant challenge for isolation [8–10]. Technological advancement has led to the development of various approaches to detect [11] and isolate CTCs [12–15]. Recent studies have found CTCs as single cells or clusters, which vary in number depending on cancer type and stage. Many CTCs are expected to be nonviable or fragile in vivo due to anoikis—cell death due to loss of attachment—or stressful environmental conditions in the blood, such as shear force. Additional CTCs may die during the process of isolation.

Despite these challenges, we have successfully established patient-derived CTC lines from luminal metastatic breast cancer patients via ex vivo culture [16]. Using these CTC lines, we have further succeeded in establishing single-cell CTC clones. Studies have shown that CTCs demonstrate inter- and intra-patient heterogeneity [17, 18], and that clonal events can drive metastasis [19]. Therefore, every viable CTC counts. Cultured single-cell CTC clones will help us understand these cells' genetic and epigenetic heterogeneity, and correlate this heterogeneity with phenotypic analysis both in vitro and in vivo. In addition, established single-cell clones will enable us to develop the right culture conditions to establish more patient-derived CTC lines. Ultimately, we may discover which of these single cell clones can create metastatic lesions—providing crucial information for guiding the clinical monitoring of cancer patients. In this protocol, we present a method to generate single cell clones from heterogeneous in vitro cultures of CTCs (Fig. 1).

2 Materials

2.1 Flow Cytometry

1% BSA in PBS buffer, keep on ice.

0.22 μ Syringe Filters, Sterile.

7-AAD 40×.

Trypan blue.

DRAQ5 (optional).

Hemocytometer.

Foil.

2.2 CTC Culture

RPMI 1640 (with phenol red) supplemented with: 20 ng/mL Recombinant Human EGF, 20 ng/mL Recombinant Human FGF-basic, 0.02% v/v B27 Supplement, 10 mg/mL Gentamicin,

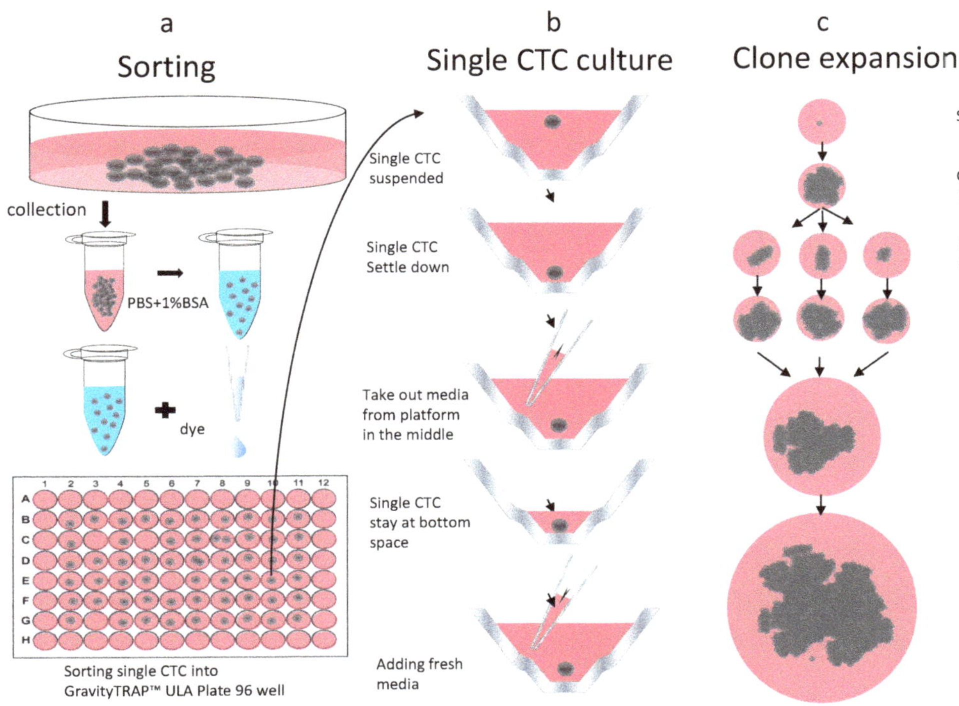

Fig. 1 Illustration of the protocol. (**a**)-(**c**) show the method of establishing single cell clones from in vitro cultured circulating tumor cells: (**a**) flow cytometry (Subheading 3.1); (**b**) single CTC culture (Subheading 3.2); and (**c**) establishment of single-cell clones (Subheading 3.2)

0.5% v/v amphotericin B/sodium deoxycholate (Gibco). Mix, shake to achieve uniformity. Keep in 4 °C (*see* **Note 1**).

Alternatively, the following media can be used:

RPMI 1640 (with phenol red) supplemented with: 20 ng/mL Recombinant Human EGF, 20 ng/mL Recombinant Human FGF-basic, 0.02% v/v B27 Supplement, 0.01% Antibiotic-Antimycotic 100×.

Mix, shake to achieve uniformity. Keep in 4 °C (*see* **Note 1**).

GravityTRAP™ ULA Plate 96 well (PerkinElmer) (*see* **Note 6**).

Ultra-low attachment 6-well plates.

Ultra-low attachment 6 cm plates.

Ultra-low attachment 24-well plates (optional).

Cell culture incubator with regulated levels of CO_2 and O_2.

2.3 Incubator Conditions for Single-Clone Cell Culture

Cells were cultured in a humid 37 C incubator with 5% CO_2 and 4% O_2. Culture media was changed every 3–4 days.

3 Methods

Keep all aforementioned materials at room temperature, unless otherwise specified.

Prepare ice.

3.1 Flow Cytometry

1. Take the CTCs cultured in suspension in ultra-low attachment plates and check the cell conditions under the microscope. Generally, cells cluster in small clumps and stay in the center of the well, and media looks clean and clear. Under the microscope, CTCs should look healthy, round, clean, and clear with a distinguishable cytoplasmic membrane, allowing light to pass through and with minimal debris in the media (Fig. 2h).
2. Transfer CTCs to a 1.7 mL sterile Eppendorf tube and centrifuge for 10–12 s (speed around 1200r cf., recommended bench mini-centrifuge; *see* **Note 2**). The cell pellet should be visible (*see* **Note 3**). Remove the supernatant slowly using a pipet.
3. Wash the cell pellet using cold sterile 1% BSA in PBS buffer, pipet up and down to mix, and centrifuge for 10–12 s (speed around 1200r cf., recommended bench mini-centrifuge). The cell pellet should be visible. Remove the supernatant slowly using a pipet (optional; *see* **Note 4**).
4. Add 1 mL of cold sterile 1% BSA in PBS buffer to Eppendorf tube, pipet up and down carefully about 100 times to fully mix, and separate cells from each other (This step will affect the sorting result.).
5. Remove the bubbles at the very top.
6. Transfer 5 μL to another Eppendorf tube, add 5 μL of Trypan blue, mix, check under the microscope using a hemocytometer, and ensure that the majority of cells are alive, healthy, and single cells.
7. Add cold 1% BSA in PBS buffer to Eppendorf tube and make the final volume 1 mL.
8. Add 7-AAD, which stains dead cells. Keep on ice, ready for sorting (*see* **Note 5**).
9. Add DRAQ5, which stains viable cells (optional; *see* **Note 5**).
10. Sort single cells into GravityTRAP™ ULA Plate 96 well (PerkinElmer) that was pre-filled with CTC media 90–100 uL/well, 1 cell per well (*see* **Notes 6–9**).

3.2 CTC Culture

1. After flow cytometry sorting, culture cells in a humid 37 °C incubator with 5% CO_2 and 4% O_2 for 2–3 h. This will allow the single cells to settle to the bottom of the well.

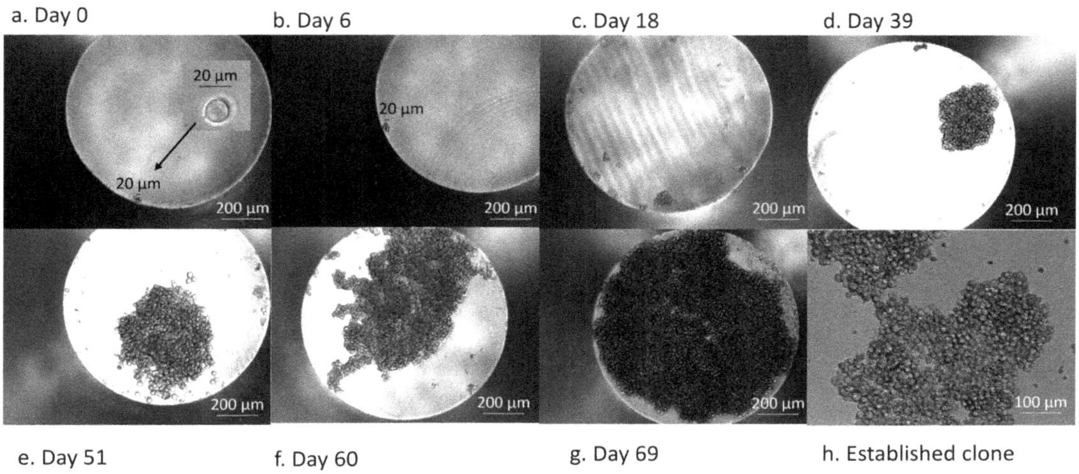

Fig. 2 Example images of an established single CTC clone at different time points. (**a**)–(**h**) show a typical cell expansion from single BRx-68 CTC under the microscope. Magnification times is 5× for (**a**)–(**g**), and 10× for (**h**). A single BRx-68 CTC in the well (**a**) immediately after sorting, and after being in culture for (**b**) 6 days, (**c**) 18 days, (**d**) 39 days, (**e**) 51 days, (**f**) 60 days, (**g**) 69 days. (**h**) A BRx-68 single clone line being moved to a 24-well plate

2. Centrifuging the plates at 200 g for 1 m, with a recommended acceleration and deceleration speed of 6, will remove bubbles and provide a clear view (*see* **Note 10**).
3. Check plates under the microscope and record cell numbers in each well, and discard wells without single cells.
4. Change media every 3–4 days using 65–75 μL/well (*see* **Notes 6** and **11**; Fig.1b), and count every 2–4 weeks. Try not to disturb the cells too much.
5. When you see the cells cover more than 80% of the well's space, move them to several wells of a 96-well plate using procedure described in Subhedading 3.1.
6. Once cells reach 80% confluence (Fig. 2g), collect them and move them into a 24-well plate, using procedure described in Subheading 3.1.
7. Repeat the **steps 5** or **6** until you get enough cells (Fig. 1c; *see* **Note 12**).

4 Notes

1. We have previously published this media for CTC culture and routine maintenance [16].
2. In this step, we find that it is better to centrifuge CTCs at a high speed for a short time (200 g, 5 min.). When using a low

speed and longer time, it is difficult to appropriately adjust the acceleration and deceleration, and the small number of cells can stick to the walls, leading to the loss of many CTCs. In addition, shorter experimental times will lead to higher survival efficiency.

3. Minimum cell number depends on your lab's flow cytometry efficiency and CTC cell culture survival rate.
4. This step will result in the loss of a lot of cells (optional).
5. Concentration of dye depends on the datasheet from respective manufacturer.
6. In this kind of plate, the well has a super tiny flat bottom to hold cells in place (diameter of bottom is 1 mm). In comparison, the diameter of a regular 96-well plate is 6.4 mm. This design facilitates finding the cells within the well under 4× or 10× magnification, monitoring them, and enhancing their survival (Fig. 2a). In the middle of each well, there is an additional platform for resting the pipet tip while changing media, in order to prevent accidentally aspirating suspended CTCs at the bottom (Fig. 1b).
7. Do this step as fast as possible, since cells can die if they spend a long time outside the incubator. If there is a long distance between the lab and sorting facility, you can wrap the plates with aluminum foil to protect them from the light and ambient oxygen.
8. Flow cytometry sorting equipment has differing efficiencies, which can significantly influence results.
9. Avoid sorting the cells into the edges of this 96-well plate, since evaporation at the edges can kill cells (optional).
10. After sorting, give the cells time (3 h or less) to naturally settle to the bottom. If you centrifuge immediately after sorting, you can potentially miss CTCs that stick to the walls of the well. If you let the cells settle for more than 3 h, single cells may proliferate, causing you to discard wells as having more than one cell.
11. Change the tip for each well in order to avoid cell contamination.
12. It is best to keep the cells at a high confluence, in order to promote survival and proliferation.

Acknowledgments

Our research is funded by the NIH K22 award (K22 CA175228-01A1), NIH DP2 award (DP2 CA206653), Donald E. and Delia B. Baxter Foundation, Stop Cancer Foundation, Wright

Foundation, the Pew Charitable Trusts, and the Alexander & Margaret Stewart Trust (Pew-Stewart Scholar for Cancer Research) to M.Y. We are grateful to Ms. C. Lytal for assisting with manuscript editing.

References

1. Allemani C, Weir HK, Carreira H, Harewood R, Spika D, Wang XS, Bannon F, Ahn JV, Johnson CJ, Bonaventure A, Marcos-Gragera R, Stiller C, Azevedo e Silva G, Chen WQ, Ogunbiyi OJ, Rachet B, Soeberg MJ, You H, Matsuda T, Bielska-Lasota M, Storm H, Tucker TC, Coleman MP, Group CW (2015) Global surveillance of cancer survival 1995–2009: analysis of individual data for 25,676,887 patients from 279 population-based registries in 67 countries (CONCORD-2). Lancet 385(9972):977–1010. https://doi.org/10.1016/S0140-6736(14)62038-9
2. Weigelt B, Peterse JL, van't Veer LJ (2005) Breast cancer metastasis: markers and models. Nat Rev Cancer 5(8):591–602. https://doi.org/10.1038/nrc1670
3. Yadavalli S, Jayaram S, Manda SS, Madugundu AK, Nayakanti DS, Tan TZ, Bhat R, Rangarajan A, Chatterjee A, Gowda H, Thiery JP, Kumar P (2017) Data-driven discovery of extravasation pathway in circulating tumor cells. Sci Rep 7:43710. https://doi.org/10.1038/srep43710
4. Aceto N, Bardia A, Miyamoto DT, Donaldson MC, Wittner BS, Spencer JA, Yu M, Pely A, Engstrom A, Zhu H, Brannigan BW, Kapur R, Stott SL, Shioda T, Ramaswamy S, Ting DT, Lin CP, Toner M, Haber DA, Maheswaran S (2014) Circulating tumor cell clusters are oligoclonal precursors of breast cancer metastasis. Cell 158(5):1110–1122. https://doi.org/10.1016/j.cell.2014.07.013
5. Baccelli I, Schneeweiss A, Riethdorf S, Stenzinger A, Schillert A, Vogel V, Klein C, Saini M, Bauerle T, Wallwiener M, Holland-Letz T, Hofner T, Sprick M, Scharpff M, Marme F, Sinn HP, Pantel K, Weichert W, Trumpp A (2013) Identification of a population of blood circulating tumor cells from breast cancer patients that initiates metastasis in a xenograft assay. Nat Biotechnol 31(6):539–544. https://doi.org/10.1038/nbt.2576
6. Hayes DF, Cristofanilli M, Budd GT, Ellis MJ, Stopeck A, Miller MC, Matera J, Allard WJ, Doyle GV, Terstappen LW (2006) Circulating tumor cells at each follow-up time point during therapy of metastatic breast cancer patients predict progression-free and overall survival. Clin Cancer Res 12(14 Pt 1):4218–4224. https://doi.org/10.1158/1078-0432.CCR-05-2821
7. Heller G, McCormack R, Kheoh T, Molina A, Smith MR, Dreicer R, Saad F, de Wit R, Aftab DT, Hirmand M, Limon A, Fizazi K, Fleisher M, de Bono JS, Scher HI (2018) Circulating tumor cell number as a response measure of prolonged survival for metastatic castration-resistant prostate cancer: a comparison with prostate-specific antigen across five randomized phase III clinical trials. J Clin Oncol 36(6):572–580. https://doi.org/10.1200/JCO.2017.75.2998
8. Khoo BL, Grenci G, Lim YB, Lee SC, Han J, Lim CT (2018) Expansion of patient-derived circulating tumor cells from liquid biopsies using a CTC microfluidic culture device. Nat Protoc 13(1):34–58. https://doi.org/10.1038/nprot.2017.125
9. Stott SL, Lee RJ, Nagrath S, Yu M, Miyamoto DT, Ulkus L, Inserra EJ, Ulman M, Springer S, Nakamura Z, Moore AL, Tsukrov DI, Kempner ME, Dahl DM, Wu CL, Iafrate AJ, Smith MR, Tompkins RG, Sequist LV, Toner M, Haber DA, Maheswaran S (2010) Isolation and characterization of circulating tumor cells from patients with localized and metastatic prostate cancer. Sci Transl Med 2(25):25ra23. https://doi.org/10.1126/scitranslmed.3000403
10. Hodgkinson CL, Morrow CJ, Li Y, Metcalf RL, Rothwell DG, Trapani F, Polanski R, Burt DJ, Simpson KL, Morris K, Pepper SD, Nonaka D, Greystoke A, Kelly P, Bola B, Krebs MG, Antonello J, Ayub M, Faulkner S, Priest L, Carter L, Tate C, Miller CJ, Blackhall F, Brady G, Dive C (2014) Tumorigenicity and genetic profiling of circulating tumor cells in small-cell lung cancer. Nat Med 20(8):897–903. https://doi.org/10.1038/nm.3600
11. USFDA (2008) CallSearch-circulating tumor cell kit: K073338
12. Karabacak NM, Spuhler PS, Fachin F, Lim EJ, Pai V, Ozkumur E, Martel JM, Kojic N, Smith K, Chen PI, Yang J, Hwang H, Morgan B, Trautwein J, Barber TA, Stott SL, Maheswaran S, Kapur R, Haber DA, Toner M (2014) Microfluidic, marker-free isolation of

circulating tumor cells from blood samples. Nat Protoc 9(3):694–710. https://doi.org/10.1038/nprot.2014.044

13. Sarioglu AF, Aceto N, Kojic N, Donaldson MC, Zeinali M, Hamza B, Engstrom A, Zhu H, Sundaresan TK, Miyamoto DT, Luo X, Bardia A, Wittner BS, Ramaswamy S, Shioda T, Ting DT, Stott SL, Kapur R, Maheswaran S, Haber DA, Toner M (2015) A microfluidic device for label-free, physical capture of circulating tumor cell clusters. Nat Methods 12(7):685–691. https://doi.org/10.1038/nmeth.3404
14. Fachin F, Spuhler P, Martel-Foley JM, Edd JF, Barber TA, Walsh J, Karabacak M, Pai V, Yu M, Smith K, Hwang H, Yang J, Shah S, Yarmush R, Sequist LV, Stott SL, Maheswaran S, Haber DA, Kapur R, Toner M (2017) Monolithic chip for high-throughput blood cell depletion to sort rare circulating tumor cells. Sci Rep 7(1):10936. https://doi.org/10.1038/s41598-017-11119-x
15. Chen L, Peng M, Li N, Song Q, Yao Y, Xu B, Liu H, Ruan P (2018) Combined use of EpCAM and FRalpha enables the high-efficiency capture of circulating tumor cells in non-small cell lung cancer. Sci Rep 8(1):1188. https://doi.org/10.1038/s41598-018-19391-1
16. Yu M, Bardia A, Aceto N, Bersani F, Madden MW, Donaldson MC, Desai R, Zhu H, Comaills V, Zheng Z, Wittner BS, Stojanov P, Brachtel E, Sgroi D, Kapur R, Shioda T, Ting DT, Ramaswamy S, Getz G, Iafrate AJ, Benes C, Toner M, Maheswaran S, Haber DA (2014) Cancer therapy. Ex vivo culture of circulating breast tumor cells for individualized testing of drug susceptibility. Science 345(6193):216–220. https://doi.org/10.1126/science.1253533
17. De Mattos-Arruda L, Cortes J, Santarpia L, Vivancos A, Tabernero J, Reis-Filho JS, Seoane J (2013) Circulating tumour cells and cell-free DNA as tools for managing breast cancer. Nat Rev Clin Oncol 10(7):377–389. https://doi.org/10.1038/nrclinonc.2013.80
18. Messaritakis I, Stoltidis D, Kotsakis A, Dermitzaki EK, Koinis F, Lagoudaki E, Koutsopoulos A, Politaki E, Apostolaki S, Souglakos J, Georgoulias V (2017) TTF-1- and/or CD56-positive circulating tumor cells in patients with small cell lung cancer (SCLC). Sci Rep 7:45351. https://doi.org/10.1038/srep45351
19. Hunter KW, Amin R, Deasy S, Ha NH, Wakefield L (2018) Genetic insights into the morass of metastatic heterogeneity. Nat Rev Cancer 18(4):211–223. https://doi.org/10.1038/nrc.2017.126

Chapter 9

Immunofluorescence Combined with Single-Molecule RNA Fluorescence In Situ Hybridization for Concurrent Detection of Proteins and Transcripts in Stress Granules

Jakub Kochan and Mateusz Wawro

Abstract

Immunofluorescence (IF) microscopy is arguably one of the most commonly used methods for studying structure and composition of stress granules (SGs). While in most cases standard IF protocols are sufficient to visualize protein components of SGs, concurrent detection of proteins and transcripts in stress granules requires more sophisticated and problematic approaches. Here we present a well-established, simple, robust, and fluorescent protein-compatible method for simultaneous detection of proteins and transcripts in individual stress granules using combination of IF and single-molecule RNA fluorescence in situ hybridization (smRNA FISH).

Key words Microscopy, Immunofluorescence, smRNA FISH, Stress granules, RNA, lncRNA, Single-cell analysis

1 Introduction

Stress granules (SGs) are large ribonucleoprotein complexes predominantly comprised of stalled translation initiation complexes, RNA-binding proteins, and signaling factors engaged in numerous aspects of cellular metabolism. These non-membrane-bound assemblies of RNAs and proteins form when cells are subjected to adverse conditions such as heat shock (e.g., hyperthermia), viral infections, oxidative stress, UV light, amino acid deprivation, or osmotic shock. SGs are dynamic structures involved in RNA triage and modulation of various signaling cascades determining whether stressed cells should eventually survive or undergo apoptosis [1, 2].

The key feature of stress granules is that virtually all known protein components found in SGs pre-exist dispersed in the cytoplasm of resting cells and rapidly and reversibly accumulate by liquid-liquid phase separation into a condensed state upon exposure to stress conditions. In this condensed phase, the RNAs and

Miodrag Gužvić (ed.), *Single Cell Analysis: Methods and Protocols*, Methods in Molecular Biology, vol. 2752, https://doi.org/10.1007/978-1-0716-3621-3_9,

RNA-binding proteins behave as a single organelle with liquid-like properties forming cytologically discernible foci [2–4].

The protein composition, structure, and dynamics of SG formation have been extensively studied over the last two decades. Due to the fact that formation of SGs is a result of subcellular re-localization of specific components rather than change in their cellular level, the method of choice in studies on SGs was and still is fluorescence microscopy (Fig. 1). This approach has provided valuable insights into the spatio-temporal organization of protein components of stress granules and shed light on the molecular mechanisms governing formation of SGs [5].

Whereas the dynamics of SGs formation and their main protein components were recently characterized [1, 6–8], the RNA composition of SGs still remains poorly explored. Only a few specific mRNAs have been shown to localize to SGs. The full population of mRNAs in SGs, and whether SGs contain any specific noncoding RNAs (ncRNAs), is not known. Only recently, in an attempt to describe the mammalian stress granule transcriptome using RNA-seq analysis, Khong et al. revealed that while essentially every mRNA can be targeted to stress granules, there are also certain lncRNAs that are overrepresented in these structures [9]. These observations along with spectacular progress in SG research led to increased interest in procedures allowing for concurrent observations of specific SG markers (proteins) and transcripts in stress granules at the level of single cells.

Though the concept of linking immunofluorescence with single-molecule RNA fluorescence in situ hybridization (smRNA FISH) seems apparent, the specific materials used during IF and smRNA FISH make this approach difficult to perform on the same specimen. Here we present a well-established, straightforward, robust, and fluorescent protein-compatible method for simultaneous detection of proteins and transcripts in individual SGs using combination of IF and smRNA FISH. The presented protocol is based on our previous seminal technical report published in 2015 [10]. To date, our procedure has been successfully used by several groups. Lately we managed to adopt the original protocol to visualize components of stress granules. Here we present protocol allowing for simultaneous observation of a key protein marker of SGs, G3BP1, and lncRNA *NORAD* recently found to be one of the top five most abundant lncRNAs in human SGs [9].

2 Materials

Read this section carefully and ensure to have all reagents prepared before beginning the procedure. Also make sure to read the detailed protocol and respective notes in advance. Observe caution not to introduce RNase contamination into your reagents and

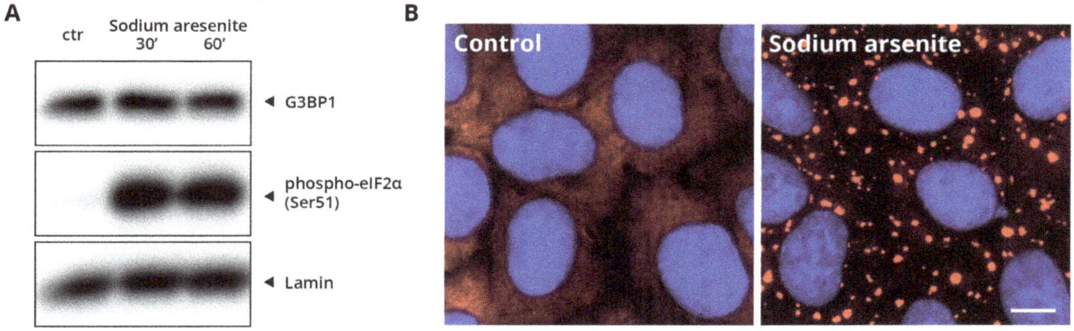

Fig. 1 Formation of stress granules is a result of subcellular re-localization of specific components rather than change in their cellular level. (**a**) Western blot analysis of extracts from HeLa cells, control or sodium arsenite-treated (500 μM, 30 and 60 min), using anti-G3BP1 (upper panel), anti-phospho-eIF2α (Ser51, middle panel), and anti-lamin (loading control, lower panel) antibodies. While no change in total G3BP level is detected upon treatment with sodium arsenite, observed phosphorylation of eIF2α clearly reveals exposure of cells to stress conditions. (**b**) Immunofluorescence-based approach allows observation of stress granules formed upon treatment of the cells with sodium arsenite. Images of HeLa cells with visualized G3BP1, a key marker of stress granules (red) in control and sodium arsenite-treated cells. Blue: DAPI (nuclei). Scale bar 10 μm

processed specimens. Wear gloves throughout the whole experiment and change them often, especially when you suspect contact with skin, hair, or other surfaces potentially contaminated with RNases (e.g., keyboards, doorknobs etc.). Use of sterile, RNase-free containers and consumables at every step is recommended (*see* **Note 1**). Prepare all solutions using ultrapure (i.e., 18.2 MΩ·cm at 25 °C) RNase-free water (*see* **Note 2**). We do not add sodium azide or other preservatives to the prepared reagents.

2.1 Cell Culture

1. HeLa cells (human cervix adenocarcinoma).
2. Dulbecco's Modified Eagle Medium (DMEM) with 1.0 g/L D-glucose (Lonza) supplemented with GlutaMAX (Gibco) and 10% fetal bovine serum (FBS, heat inactivated, South America origin, Biowest).
3. Phosphate buffered saline (PBS, without calcium and magnesium).
4. TrypLE Express dissociating reagent (Gibco). No antibiotics (i.e., penicillin or streptomycin) were used. Cells were maintained in cell culture vessels from BD Falcon.
5. Sterile glass coverslips (*see* **Note 3**).
6. Sterile tweezers (*see* **Note 4**).

2.2 Fixation/Post-fixation

1. RNase-free Phosphate-Buffered Saline (PBS), 1× sterile: 137 mM NaCl, 2.7 mM KCl, 8 mM Na_2HPO_4, and 2 mM KH_2PO_4. Prepare by diluting Ambion 10× RNase-free PBS pH 7.4 (Invitrogen) to 1× solution in UltraPure DNase/

RNase-Free Distilled Water (Invitrogen) or equivalent (*see* **Note 2**). Store at room temperature (RT).

2. Fixative: 4% (w/v) methanol-free formaldehyde in 1× RNase-free PBS. In a 50 mL sterile, RNase-free conical tube (or appropriate sterile, RNase-free glass bottle), mix 10 mL of 16% (w/v) methanol-free formaldehyde (Invitrogen; use whole ampule at once), 4 mL of 10× RNase-free PBS pH 7.4 (Invitrogen), and 26 mL of UltraPure DNase/RNase-Free Distilled Water (Invitrogen) or equivalent. Store at 4 °C in a container wrapped with aluminum foil.

2.3 Immunofluorescence

1. RNase-free PBS, 1× sterile: see previous section.
2. 10% (v/v) Triton X-100 in RNase-free water: mix 5 mL of molecular biology-grade Triton X-100 and 45 mL of UltraPure™ DNase/RNase-Free Distilled Water (Invitrogen) or equivalent. Sterile filter through a 0.2 μm syringe filter. Store at 4 °C (*see* **Note 5**).
3. 200 mM vanadyl ribonucleoside complexes (VRC): we use molecular biology-grade VRC (Merck/Sigma-Aldrich, product number 94740). Reconstitute the powder in sterile, RNase-free water to a green-black clear solution by incubating securely sealed vial at 65 °C (about 10 min with occasional mixing). Once prepared, aliquot the entire volume into smaller samples and keep frozen at −20 °C. After thawing ensure that the solution is clear. If a visible precipitate formed, re-heat the aliquot at 65 °C.
4. 10% (w/v) sterile-filtered, RNase-free (acetylated) bovine serum albumin (Ac-BSA). We prepare Ac-BSA according to the protocol "Removal of RNase from Bovine Serum Albumin by Acetylation Reaction" available online from web resources of Malaysian Cocoa Board (http://cocoabiotech.koko.gov.my/RNases%20Elimination1.html). The protocol can be scaled-up to at least 200 mL. After dialysis against sterile, RNase-free water, we concentrate the obtained Ac-BSA using Amicon Ultra centrifugal filters (Ultracel 50 K) in refrigerated centrifuge (4 °C) equipped with swing-bucket rotor capable of accommodating 50 mL tubes to obtain 10% w/v Ac-BSA. Determine the final protein concentration using BCA assay. Filter sterilize through a 0.2 μm filter and store 2 mL aliquots at −20 °C (*see* **Notes 6** and **7**).
5. Blocking/permeabilizing buffer: 2 mM VRC, 1% (w/v) Ac-BSA, 0.3% (v/v) Triton X-100, 1× sterile, RNase-free PBS. Prepare in a 50 mL sterile, RNase-free conical tube (or appropriate sterile, RNase-free glass bottle). For 20 mL, mix 200 μl of 200 mM VRC, 2 mL of 10% Ac-BSA, 608 μl of 10% (v/v) Triton X-100, 2 mL of sterile 10× RNase-free

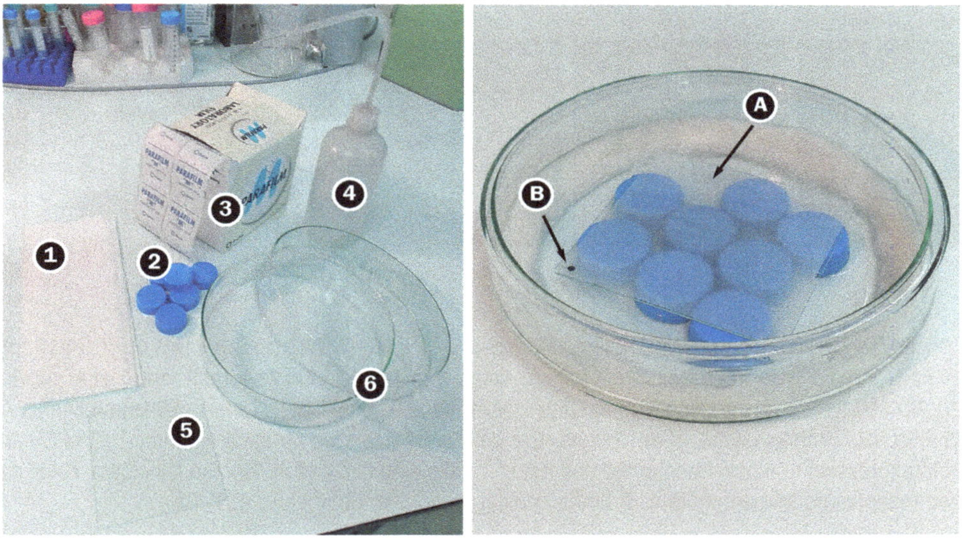

Fig. 2 The humidifying chamber assembly. Components used to build the humidifying chamber are presented in the left picture: (1) paper towels, (2) 50 mL conical tube screws, (3) parafilm, (4) distilled water, (5) glass plate, and (6) glass Petri dish. Right image presents a typical view of an assembled chamber (**a** – a glass plate with piece of parafilm placed on top (stage for coverslips), (**b**) – an orientation mark on the parafilm used for proper alignment of coverslips)

PBS, and add sterile, RNase-free water to 20 mL. Sterile filter through a 0.2 μm syringe filter. Store at 4 °C in a container wrapped with aluminum foil.

6. Primary and secondary antibodies. We use mouse monoclonal antibodies anti-human G3BP1 (250 μg/mL, BD Transduction Laboratories, 611126), diluted 1:50 and goat anti-mouse IgG (H + L) highly cross-adsorbed secondary antibodies conjugated with Alexa Fluor 488 (2 mg/mL, Invitrogen, A-11029), diluted 1:500).
7. Humidifying chamber (Fig. 2).

2.4 smRNA FISH

1. 20× SSC buffer: 3 M sodium chloride, 300 mM sodium citrate. We use Ambion 20× SSC solution (Invitrogen). Store at RT. 20× SSC can be also easily prepared in house (*see* **Note 8**).
2. 2× SSC buffer. Prepare in a 50 mL sterile, RNase-free conical tube. For 50 mL mix 5 mL 20× SSC and 45 sterile, RNase-free water. Store at RT.
3. Wash buffer: 2× SSC, 10% (v/v) formamide. Prepare in a 50 mL sterile, RNase-free conical tube. For 50 mL, mix 5 mL 20× SSC, 5 mL formamide, and add sterile, RNase-free water to 50 mL (*see* **Note 9**). Sterile filter through a 0.2 μm syringe filter. Store at RT.

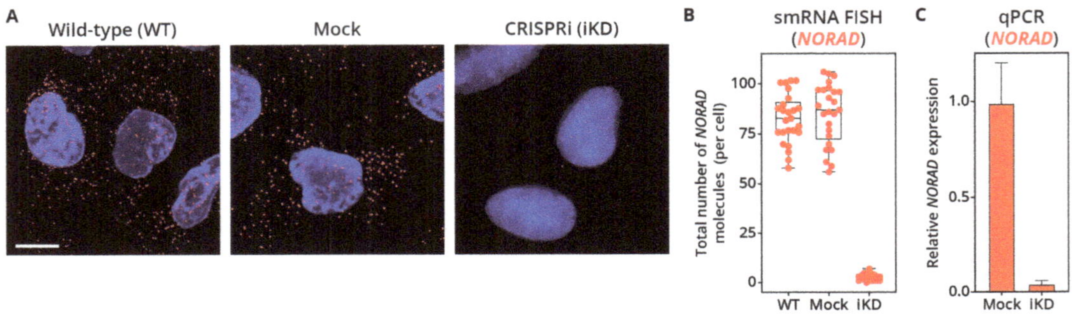

Fig. 3 Specificity verification of used probe blend targeting lncRNA *NORAD*. (**a**) Detection of *NORAD* (red) using smRNA FISH in wild-type HeLa cells (left panel), mock-transfected (middle panel), and $HeLa^{iKD}$ cells with induced *NORAD* knock-down (right panel). Blue: DAPI (nuclei). Scale bar: 10 μm (**b**) Quantification of the smRNA FISH experiments shown in (**a**). Twenty-five cells were analyzed for each condition. (C) qPCR analysis of *NORAD* expression in mock-transfected and $HeLa^{iKD}$ cells with induced *NORAD* knock-down. Bars: average from two independent experiments with two technical replicates. Error bars: 95% CI

4. Hybridization buffer: 2× SSC, 10% (v/v) formamide, 10% (w/v) dextran sulfate. Prepare in a 50 mL sterile, RNase-free conical tube. For 10 mL, mix 1 mL of 20× SSC, 1 mL formamide, 1 g of dextran sulfate, and add sterile, RNase-free water to 10 mL. Store 1 mL aliquots at −20 °C (*see* **Note 10**).
5. smRNA FISH probe blend. We used custom-designed probe blend labelled with Quasar 570 dye (LGC Biosearch Technologies) targeting lncRNA *NORAD* at the concentration of 125 nM (Fig. 3). Probes were designed using the Stellaris Probe Designer (version 4.2) available online free of charge (www.biosearchtech.com).

2.5 Imaging

In principle every modern fluorescence microscope can be used to image prepared specimens; therefore, we suggest to consult your local microscopy facility operators for possible options. All images presented in this chapter were obtained using a Leica DM6B upright widefield fluorescence microscope (Leica Microsystems). We used a 63× oil immersion objective and a 12-bit Leica DMC5400 CMOS camera (Leica Microsystems) with Leica LAS X image acquisition software. The following filter sets (Leica Microsystems) were used: A4 for detection of DAPI (nuclei), GFP ET for detection of Alexa Fluor 488 (IF signal), and RHOD ET for Quasar 570 dye (smRNA FISH signal). After acquisition of about 30 z-sections with 0.3 μm spacing, images were deconvolved using Huygens Professional Software (Scientific Volume Imaging). Final image adjustments were performed using ImageJ 1.53c (National Institutes of Health) and Adobe Illustrator 2020 (Adobe Systems).

3 Methods

The presented protocol, including all reagent volumes, is optimized for 12-well plate format but of course it can be scaled up or down according to specific needs. The procedure usually requires *4–5 consecutive days of work*. Remember to *book a proper fluorescence microscope in advance* as imaging freshly prepared specimens gives the best quality of results. *It is not recommended to store the specimens longer then overnight.* If longer storage of prepared specimens is needed, diligently follow the storage instructions provided by the manufacturer of the used mounting medium.

3.1 Plating and Treatment of the Cells

1. Using tweezers place sterile glass coverslips in a multi-well cell culture plate and seed the cells at the desired density suitable for single-cell imaging and analysis. Usually 50,000–60,000 cells per well of 12-well plate in 1 mL of medium should be sufficient. In general, it is not recommended to let the cells reach 100% confluence (i.e., monolayer) at the time of fixation (*see* **Note 11**).
2. Place the cells in a dedicated incubator for 24–48 h to allow proper attachment and accommodation. If needed, change the medium at desired time point or treat the cells according to your experimental requirements.
3. Induce stress granule formation (*see* **Note 12**) and proceed immediately to subsequent steps.

3.2 Fixation

1. Using a sterile, RNase-free Pasteur pipette connected to a vacuum pump, remove and dispose properly the culture medium.
2. Wash the cells twice with 1× PBS (1 mL/well) (*see* **Note 13**). Use a serological RNase-free pipette. Work gently and fluently. Do not let the cells to dry out.
3. Remove PBS, add 1 mL of fixative, and incubate the cells for 10–15 min at room temperature (*see* **Note 14**).
4. Remove fixative and wash the cells 3 times with 1× PBS (1 mL/well, 5 min at room temperature). Gentle agitation using an orbital shaker will facilitate proper washing (*see* **Note 15**).

3.3 Immunofluorescence

1. Following fixation, remove PBS and block/permeabilize the cells for 60 min at RT with gentle shaking using 1 mL/well of blocking buffer (*see* **Note 16**).
2. While the cells are being blocked/permeabilized, prepare primary antibody by diluting them in blocking buffer. When calculating the volume of the needed antibody solution, take 50 μl of antibody solution per specimen (coverslip) and add

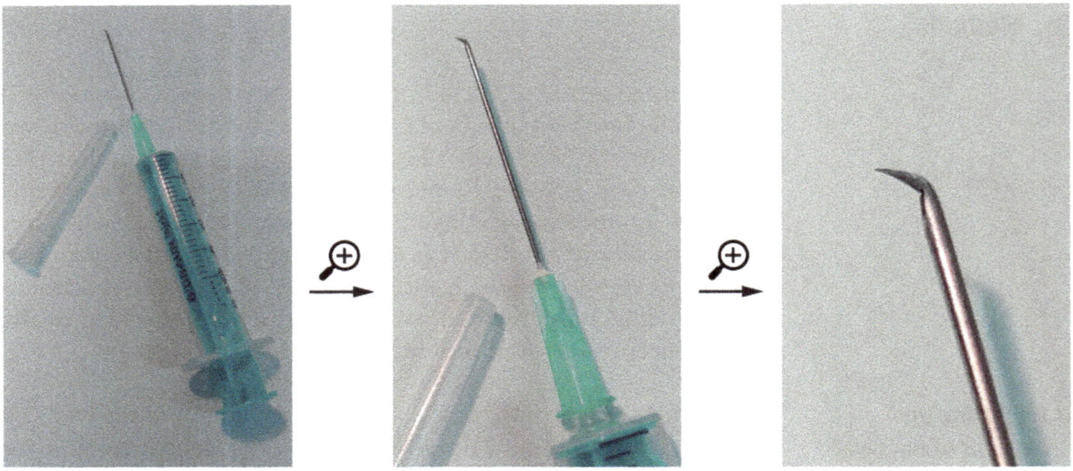

Fig. 4 Picture of the bent needle attached to a syringe used to manipulate coverslips. Zooms present the tip of the bent needle in more details

50 µl of excess to compensate for possible liquid losses and pipetting errors.

3. Assemble the humidifying chamber omitting the parafilm-coated glass plate and equilibrate it at 4 °C (see Fig. 2).
4. Carefully place 50 µl droplets of primary antibody solution on the parafilm. Avoid any air bubbles in the droplets as the cells will not be stained in these areas. While placing the droplets pay attention to arrange them in a manner allowing to place the coverslips separately and not touching each other.
5. Place the coverslips (*cells toward the liquid*) using tweezers and a bent needle (fresh, sterile, RNase-free) attached to a syringe on the primary antibody droplets (Figs. 4 and 5). *Do not squeeze the cells.*
6. Place the glass plate with specimens in humidifying chamber and incubate the cells with primary antibodies overnight at 4 °C.
7. Following the overnight incubation with primary antibodies, gently transfer the specimens into a fresh, sterile cell culture plate wells filled with 1× PBS (*rotate the coverslips so the cells are facing upwards*). Again, be careful not to disturb or dry the cells.
8. Wash the cells three times with 1× PBS (5 min at RT with gentle agitation each).
9. While washing the specimens, equilibrate the humidifying chamber to RT and set up a glass plate with a new piece of parafilm placed on top.

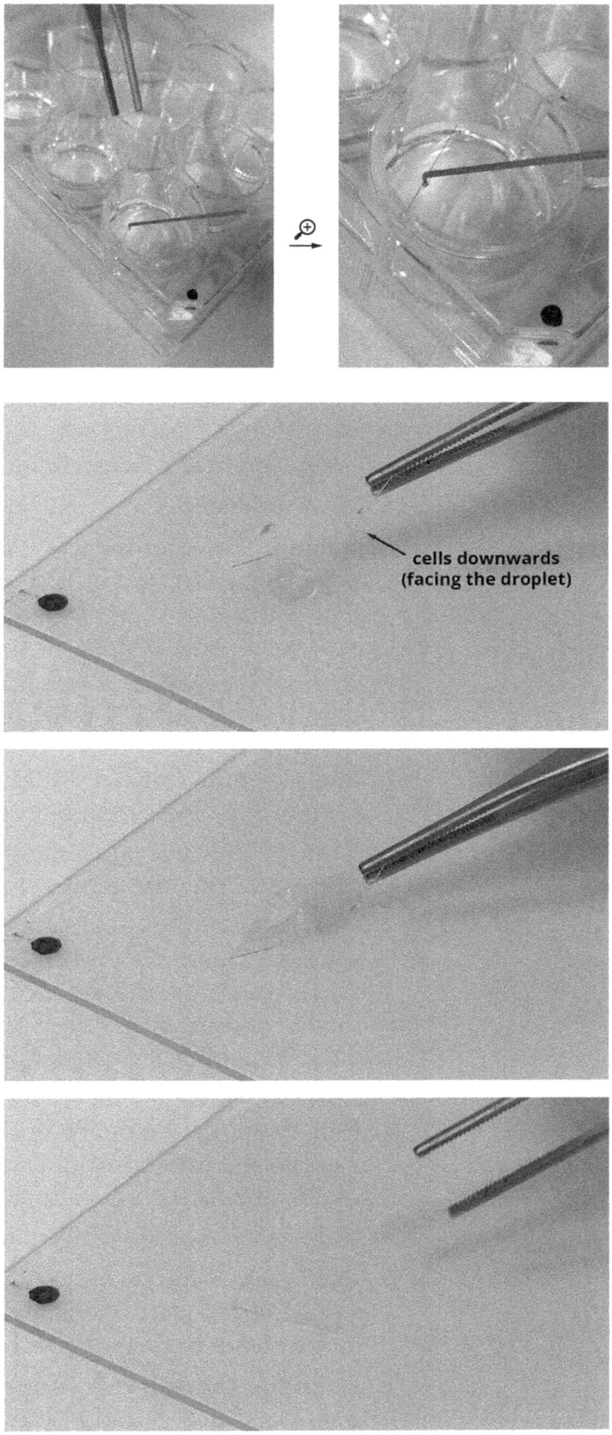

Fig. 5 Photographs presenting the essential steps during coverslip manipulation. Upper panel shows technique used to lift the coverslip from cell-culture plate well. Below a series of images presents how to place the coverslip (*cells toward the droplet*) using tweezers on the antibody/probe droplet

10. Prepare secondary antibody dilution in blocking buffer (volume of secondary antibodies should be calculated similarly to the volume of primary antibodies as in **step 2**).
11. Place 50 μl droplets of secondary antibodies on the parafilm.
12. Use tweezers and a bent needle (a new one) attached to a syringe to place coverslips (*remember to keep the cells towards the droplet*) onto the secondary antibody droplets.
13. Gently transfer the glass plate with coverslips into the humidifying chamber and incubate the specimens with secondary antibodies for 90 min at RT, in the dark (*see* **Note 17**).
14. Following the incubation with secondary antibodies, transfer the specimens into a fresh, sterile cell culture plate wells filled with 1× PBS (*remember to rotate the coverslips and keep the cells facing upwards*). Be careful not to disturb or dry the cells.
15. Wash as in **step 8**.
16. Remove PBS after last wash, add 1 mL of fixative (4% formaldehyde in PBS), and incubate for 10 min at RT (in the dark). *The post-fixation step prevents removal of antibodies during smRNA FISH steps.*
17. Remove the fixative and wash as in **step 8**. Keep the cells in PBS until starting smRNA FISH procedure.

3.4 Single-Molecule RNA Fluorescence In Situ Hybridization

1. Prepare fresh smRNA FISH wash buffer and thaw a fresh aliquot of hybridization buffer.
2. Remove PBS and equilibrate the specimens by incubation in smRNA FISH wash buffer (1 mL/well) for 5–10 min at RT.
3. During specimen equilibration, prepare probe blend working solution. Dilute the probe blend stock in hybridization buffer. When calculating the volume of the probe, you need to follow a similar routine as for the primary and secondary antibodies. Take 50 μl of probe droplet per coverslip and add 50 μl excess to compensate for liquid losses.
4. Equilibrate the humidifying chamber to *37 °C* and prepare a glass plate with a piece of parafilm placed on it (use the same chamber construction as in previous steps).
5. Place droplets of probe blend (50 μl/coverslip) on the parafilm.
6. Use tweezers and a bent needle (yet another new one) attached to a syringe to place coverslips (*cells toward the droplet*) onto the probe blend droplets (*see* **Note 18**).
7. Hybridize the cells with probes at *37 °C* overnight in the dark.
8. Following the overnight hybridization, gently transfer the specimens into a fresh, sterile cell culture plate wells filled with smRNA FISH wash buffer prewarmed to *37 °C* (*rotate the*

coverslips so the cells are facing upwards). Again, be careful not to damage the cells (*see* **Note 19**).

9. Wash the specimens twice with 1 mL of smRNA FISH wash buffer per coverslip (30 min at *37°C*, agitation is not required). Add nuclear counterstain (e.g., DAPI or equivalent) to the smRNA FISH wash buffer during the second wash (*see* **Note 20**).
10. Remove smRNA FISH wash buffer and immerse the specimens in 2× SSC while preparing to mount the coverslips onto microscope slides.
11. Mount the specimens in an appropriate fluorescence microscopy mounting medium and proceed to imaging.

3.5 Imaging

We recommend acquiring three-dimensional stacks of images using widefield fluorescence microscope. A good starting point is a 1 s or longer exposure time to detect the smRNA FISH signal. Often we use 1500 ms exposure time when working with probe blends labeled with Quasar 570 Dye. Acquire stacks of about 30 Z-slices at intervals of 300 nm in the Z direction that can be then reconstructed (deconvolved) using appropriate software (we use Huygens Professional Software). We strongly advice to test different deconvolution settings at initial stages of your experiments as it may differ between used microscopy platforms. It is not uncommon to be unable to see separated RNA spots before deconvolution (Fig. 6). Figure 7 presents anticipated results of the described procedure.

4 Notes

1. In most cases disposable cell culture plasticware is certified to be RNase-free. To remove RNases from glassware or metalware bake them at 300 °C for 4 h or 180 °C or higher for several hours. Alternatively, small metalware (e.g., spoons or spatulas) can be quickly decontaminated by holding them in a Bunsen burner flame for several seconds.
2. In most cases we use UltraPure DNase/Rnase-Free Distilled Water purchased from Invitrogen. However, when larger volumes are needed, we prefer to prepare sterile, RNase-free water in-house by diethyl pyrocarbonate (DEPC) treatment. To prepare DEPC-treated water, add DEPC to ultrapure water to a final concentration of 0.1% (v/v), i.e., 500 μl DEPC per 500 mL of water, mix the solutions thoroughly to disperse the DEPC droplets in whole volume, place the bottles in an incubator shaker providing further continuous mixing overnight,

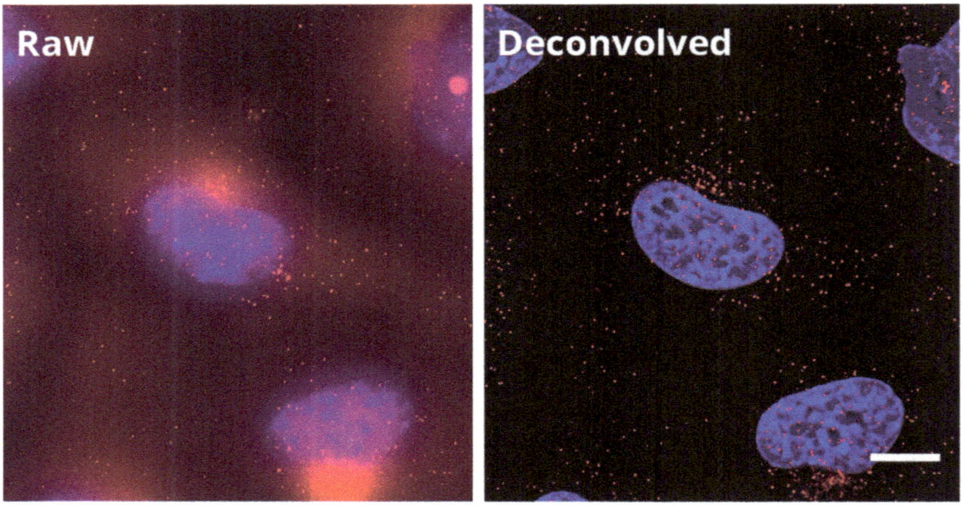

Fig. 6 Comparison of smRNA FISH images (maximum intensity projections) before (left panel) and after deconvolution (right panel). Scale bar: 10 μm

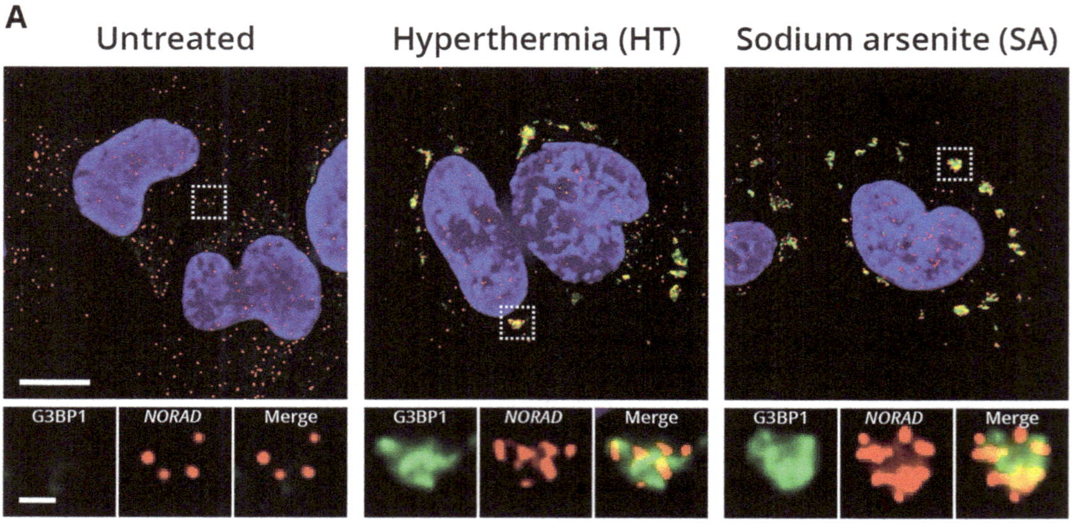

Fig. 7 Hyperthermia and sodium arsenite induce stress granule formation and lead to sequestration of lncRNA *NORAD* in stress granules. IF-smRNA FISH images of HeLa cells stained to visualize *NORAD* (red) and G3BP1, a key marker of SGs (green), during two different types of stresses (hyperthermia and sodium arsenite). Blue: DAPI (nuclei). Scale bars: 10 μm (upper panels), 1 μm (lower panels)

and then autoclave to remove the DEPC. Store at room temperature. Use sterilely.

3. Choice of shape and dimensions of coverslips depends on the cell culture plate format to be used and specific experimental requirements. As standard, for 12-well plate, we use 15 mM × 15 mM glass coverslips (No. 1.5H (170 ± 5 μm),

D 263 M Schott borosilicate glass, Menzel-Gläser). Sterilize the coverslips in a glass Petri dish by baking for at least 2 h at 180 °C.

4. Clean the tweezers using a regular detergent followed with extensive wash with tap water, distilled (demi) water, and eventually with 96% ethanol. Next, place the precleaned tweezers in the laminar flow hood and either burn them using Bunsen burner (remember to let it cool down before use) or UV-sterilize. Sterile tweezers may be kept on a freshly opened Petri dish or cell culture dish until needed.
5. Preparation of 10% (v/v) Triton X-100 may take a long while. While preparing mix the contents gently to avoid foaming. Prepare in advance and store at 4 °C.
6. For determination of Ac-BSA concentration using BCA assay, an Ac-BSA standard *must* be used. Due to severe modification after acetic anhydride treatment, normal BSA standards cannot be used. Aliquots of Ac-BSA with determined concentration may be found as components of restriction enzyme kits or purchased separately. We can also provide small aliquots of our own Ac-BSA.
7. Preparation of Ac-BSA requires use of acetic anhydride which is a restricted reagent. Because of its use in the synthesis of heroin, acetic anhydride is listed as a drug precursor and its purchase is strictly controlled. Special permissions may be required.
8. To prepare 50 mL of 20× SSC buffer, dissolve 8.765 g of sodium chloride and 1.410 g of trisodium citrate (dihydrate) in sterile, RNase-free water. Bring total volume to 50 mL.
9. Formamide is a known teratogen and should be used in a chemical fume hood only. Once opened, the bottle should be purged with inert gas (we use argon or nitrogen) and stored frozen to prevent oxidation. The breakdown products of formamide may degrade nucleic acids. Prepare fresh wash buffer for each experiment and only the volume that is needed. Use right after preparation.
10. We use dextran sulfate sodium salt from *Leuconostoc* spp. (Merck, D8906). When preparing hybridization buffer, add dextran sulfate as the last component and not at once but part after part to prevent clog formation. It may take up to 45 min. Mix gently. The hybridization buffer is very viscous. We recommend preparation of the hybridization buffer in bulk (10 mL) immediately after opening a fresh bottle of formamide and store 1 mL aliquots of the prepared buffer at −20 °C until use. This avoids any freeze-thaw effects and offers fresh hybridization buffer for each experiment. We have not done

extensive stability studies but, from our experience, the aliquots will perform to expectation for 2–3 months when stored frozen.

11. Be aware that the exact culture conditions depend on the type of used cells and desired experiment conditions. In this procedure, we used HeLa cells that usually do not require any special treatment, but some cell lines do not attach to bare (i.e., uncoated) glass coverslips and a cell line-specific treatment of the surface may be required (e.g., coating of the coverslips with poly-L-lysine, fibronectin, laminin, or other agents).
12. In the presented protocol, we induced formation of stress granules using the most commonly applied stressors: hyperthermia (42 °C, 1 h) or sodium arsenite (500 μM, 1 h). Wear protective clothing, eyewear, and gloves when handling sodium arsenite. Discard all excess sodium arsenite and media containing sodium arsenite reagents according to your institution's guidelines and all applicable local requirements.
13. From now on observe extreme caution not to introduce any RNase contamination into the specimens. All used reagents must be RNase-free.
14. Do not discard the remaining fixative. It will be used in postfixation step. Store the fixative at RT, in the dark (we wrap it in aluminum foil).
15. Possible rest point. From our experience, at this step the specimens may be kept in RNase-free PBS for some time (several hours) but we strongly recommended to process the specimens immediately after fixation to obtain best results. Do not store them overnight. In our hands the specimens are kept in RNase-free PBS at room temperature only for the time needed for preparation of blocking buffer.
16. Remember not to discard the remaining blocking buffer. It will be used to dilute primary and secondary antibodies. Store the buffer at 4 °C, in the dark (we wrap it in aluminum foil).
17. From this step onward, the specimens must be protected from light. Wrap the plate in which the specimens are washed in aluminum foil.
18. Droplets of probe in hybridization buffer exhibit different behavior compared to droplets of antibodies. Due to the presence of formamide and dextran sulfate, the hybridization buffer is more viscous and the coverslips seem not to touch fully the parafilm. It may give an impression of a stack of parafilm, buffer and coverslip with clearly distinguishable layers.
19. After overnight incubation with probe blend in hybridization buffer, the coverslips may stick to the parafilm. Observe

extreme caution while levering them with needle and forceps. Use gentle and smooth move. When placed in wash buffer, the coverslips may float on the surface of the liquid. Ensure they are covered with the wash buffer and sunk to the bottom of the well.

20. In our hands addition of nuclear counterstain to the wash buffer during second wash instead of using mounting medium with counterstain results in more even staining of nuclei and prevents the halo-like effect when imaging nuclei.

Acknowledgement

This work was supported by grant from National Science Center, Poland (project number 2020/39/D/NZ3/02328 to J.K).

References

1. Panas MD, Ivanov P, Anderson P (2016) Mechanistic insights into mammalian stress granule dynamics. J Cell Biol 215(3): 313–323. https://doi.org/10.1083/jcb.201609081
2. Protter DSW, Parker R (2016) Principles and properties of stress granules. Trends Cell Biol 26(9):668–679. https://doi.org/10.1016/j.tcb.2016.05.004
3. Molliex A, Temirov J, Lee J, Coughlin M, Kanagaraj AP, Kim HJ, Mittag T, Taylor JP (2015) Phase separation by low complexity domains promotes stress granule assembly and drives pathological fibrillization. Cell 163(1): 123–133. https://doi.org/10.1016/j.cell.2015.09.015
4. Perez-Pepe M, Fernandez-Alvarez AJ, Boccaccio GL (2018) Life and work of stress granules and processing bodies: new insights into their formation and function. Biochemistry 57(17): 2488–2498. https://doi.org/10.1021/acs.biochem.8b00025
5. Van Treeck B, Parker R (2019) Principles of stress granules revealed by imaging approaches. Cold Spring Harb Perspect Biol 11(2). https://doi.org/10.1101/cshperspect.a033068
6. Guillen-Boixet J, Kopach A, Holehouse AS, Wittmann S, Jahnel M, Schlussler R, Kim K, Trussina I, Wang J, Mateju D, Poser I, Maharana S, Ruer-Gruss M, Richter D, Zhang X, Chang YT, Guck J, Honigmann A, Mahamid J, Hyman AA, Pappu RV, Alberti S, Franzmann TM (2020) RNA-induced conformational switching and clustering of G3BP drive stress granule assembly by condensation. Cell 181(2):346–361 e317. https://doi.org/10.1016/j.cell.2020.03.049
7. Sanders DW, Kedersha N, Lee DSW, Strom AR, Drake V, Riback JA, Bracha D, Eeftens JM, Iwanicki A, Wang A, Wei MT, Whitney G, Lyons SM, Anderson P, Jacobs WM, Ivanov P, Brangwynne CP (2020) Competing protein-RNA interaction networks control multiphase intracellular organization. Cell 181(2):306–324 e328. https://doi.org/10.1016/j.cell.2020.03.050
8. Yang P, Mathieu C, Kolaitis RM, Zhang P, Messing J, Yurtsever U, Yang Z, Wu J, Li Y, Pan Q, Yu J, Martin EW, Mittag T, Kim HJ, Taylor JP (2020) G3BP1 is a tunable switch that triggers phase separation to assemble stress granules. Cell 181(2):325–345 e328. https://doi.org/10.1016/j.cell.2020.03.046
9. Khong A, Matheny T, Jain S, Mitchell SF, Wheeler JR, Parker R (2017) The stress granule transcriptome reveals principles of mRNA accumulation in stress granules. Mol Cell 68(4):808–820 e805. https://doi.org/10.1016/j.molcel.2017.10.015
10. Kochan J, Wawro M, Kasza A (2015) Simultaneous detection of mRNA and protein in single cells using immunofluorescence-combined single-molecule RNA FISH. Biotechniques 59(4):209–212, 214, 216 passim. https://doi.org/10.2144/000114340

Chapter 10

Highly Multiplexed and Simultaneous Characterization of Protein and RNA in Single Cells by Flow or Mass Cytometry Platforms Using Proximity Ligation Assay for RNA

Andrew D. Duckworth, Joseph R. Slupsky, and Nagesh Kalakonda

Abstract

In situ hybridization of oligonucleotide probes to intracellular RNA allows quantification of predefined gene transcripts within millions of single cells using cytometry platforms. Previous methods have been hindered by the number of RNA that can be analyzed simultaneously. Here we describe a method called proximity ligation assay for RNA (PLAYR) that permits highly multiplexed RNA analysis that can be combined with antibody staining. Potentially any number of RNA combined with antigen can be analyzed together, being limited only by the number of analytes that can be measured simultaneously.

Key words Transcriptome analysis, Flow cytometry, Single-cell analysis, Single-cell multi-omics, In situ hybridization, Mass spectrometry, Proteome analysis, Proximity ligation assay for RNA (PLAYR)

1 Introduction

Single-cell RNA analysis can currently be performed using two main complimentary methods of sequencing or in situ hybridization. Reverse transcription of RNA to DNA and its sequencing using high-throughput platforms can provide transcriptome-level information in single cells, yet a trade-off is ultimately observed between the sensitivity of the procedure and the number of cells analyzed due to sequencing depth and the resulting high financial cost. In situ hybridization on the other hand provides sensitive analysis of predefined RNA targets in hundreds of cells per second using cytometry-based platforms. In situ RNA hybridization methods usually contain three main steps: firstly, transcripts are specifically targeted using oligonucleotide probes; secondly, a transcript-associated sequence is replicated to amplify the signal; thirdly, this replicated transcript-specific sequence is then labeled using a complimentary DNA detection probe (DDP) that is labeled with

Miodrag Gužvić (ed.), *Single Cell Analysis: Methods and Protocols*, Methods in Molecular Biology, vol. 2752, https://doi.org/10.1007/978-1-0716-3621-3_10,

analyte. Previous methods [1–8] have been limited in the specificity as well as the number of RNA that can be simultaneously analyzed.

Here we describe a proximity ligation assay for RNA (PLAYR) [9, 10], an in situ single-cell RNA quantification technique that removes multiplexing constraints and greatly minimizes off-target signals. In addition, as PLAYR is performed at physiological conditions, the assay can be combined with the study of antigen expression using conventional antibody staining. Specific RNA detection with minimal background staining by PLAYR is facilitated through a method of proximal ligation [11, 12]. Paired oligonucleotide probes (probe pairs) are designed to anneal adjacently along their target transcript within fixed and permeabilized cells via their 5′ DNA sequence. We recommend using four to six probe pairs along the length of each target RNA to provide sufficient assay sensitivity. The number used will depend upon the copy number of the specific transcript within each cell and the efficiency of binding for each probe pair. The requirement of multiple probe pairs per transcript to maintain sensitivity of the assay usually limits PLAYR to measuring RNA which have sufficient length (e.g., mRNA, lncRNA) and viable sequence composition (i.e., sufficient GC content) to accommodate binding of multiple probe pairs. When the probe pairs are annealed and positioned alongside one another on their target RNA, their 3′ sequences combine to produce a platform for the circularization of two further DNA oligonucleotides termed the 'insert' and 'backbone'. The insert sequence is specific for each transcript while the backbone is universal. Once annealed to both probe pairs the insert and backbone form a single-stranded DNA circle that can be ligated. This completed circular DNA molecule is then copied using rolling circle amplification (RCA) which is catalyzed by the Phi29 polymerase enzyme [13]. An illustration of how the PLAYR oligonucleotides bind together is shown in Fig. 1. We recommend 2 h and 4 h RCA times for flow and mass cytometry, respectively. Increasing the RCA time beyond this can increase the signal but not generally the sensitivity of the assay (i.e., negative cells remain negative). DNA amplification during RCA initiates from the 3′ end of one of the probe pairs, producing a linear DNA molecule that is composed of potentially hundreds of concatenated complimentary copies of the original rolling circle. Thus, transcript-specific complimentary insert sequences, which have been replicated during the RCA, can then be detected using DDPs which are conjugated to a specific analyte of your choice.

PLAYR is particularly powerful technique when combined with mass cytometry [14, 15] (PLAYR-CyTOF), where up to 50 analytes can be concurrently measured. To this end, we present an additional sub-method for labeling DDPs with metal analytes for use in mass cytometry in **Methods** Subheading 3.1 of this protocol. In addition, cells analyzed by PLAYR-CyTOF may be pooled using live-cell barcoding (LCB) techniques [16–18], in which

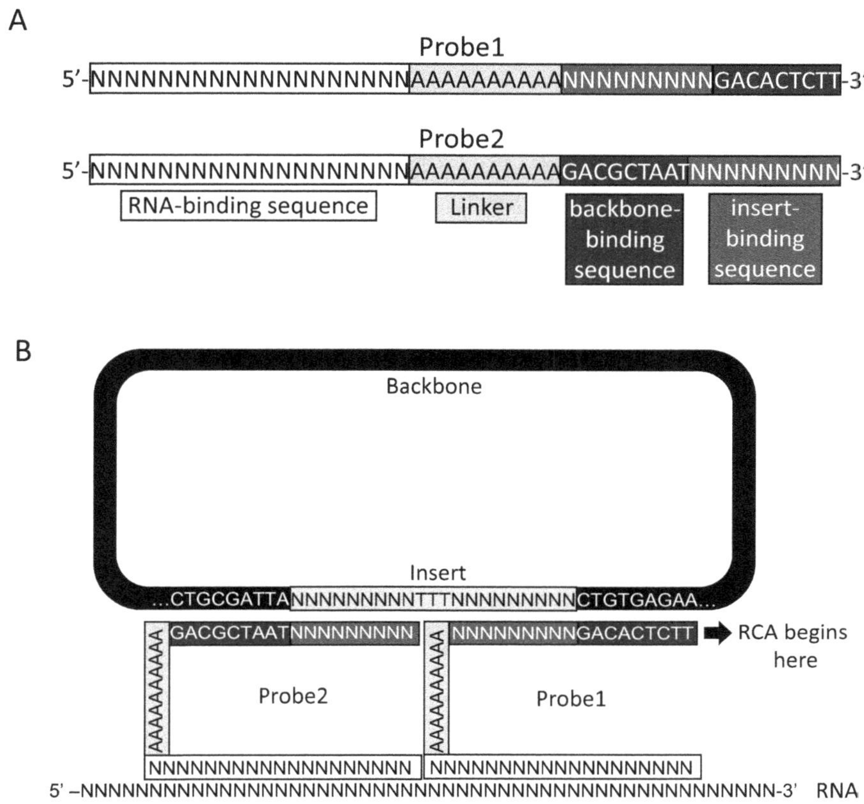

Fig. 1 Overview of PLAYR probe design and cognate interactions between oligonucleotides within the PLAYR systems. (**a**) Overview of the PLAYR probe design, with each sequence component shaded differently. (**b**) Diagram showing sequence interactions between different oligonucleotides during PLAYR. "N" represents nucleotides that vary depending upon the target (RNA-binding sequence) and the PLAYR system (insert-binding sequence) used

ubiquitously expressed abundant antigen(s) are targeted to label each sample with different combinations of heavy-metal isotopes. See Table 1 for an example of an LCB scheme using Cadmium-conjugated antibodies for barcoding of up to 20 samples.

There are currently 27 sets of PLAYR systems (i.e., corresponding probe pair 3′ ends, inserts, backbones, and DDPs) each allowing simultaneous quantification of a transcript. All 27 PLAYR system sequences are provided (Table 2), with room for expansion if required. In addition, we provide validated predesigned 5′ probe pair sequences for targeting common "housekeeping" transcripts (Table 3). Instructions on installing and using PLAYRDesign software, an R software which aids in picking 5′ probe pair sequences and designing PLAYR probes for your RNA target of interest, is available at https://github.com/nolanlab/PLAYRDesign. Specificity and accurate quantification of transcripts using designed PLAYR probes should be validated by

Table 1
Example of a 6-choose-3 LCB scheme using Cadmium isotopes

		Cadmium Isotopes					
		106	110	111	113	114	116
Barcode	1	x	x	x			
	2	x	x		x		
	3	x	x			x	
	4	x	x				x
	5	x		x	x		
	6	x		x		x	
	7	x		x			x
	8	x			x	x	
	9	x			x		x
	10	x				x	x
	11		x	x	x		
	12		x	x		x	
	13		x	x			x
	14		x		x	x	
	15		x		x		x
	16		x			x	x
	17			x	x	x	
	18			x	x		x
	19			x		x	x
	20				x	x	x

An antibody (or antibodies) that target constituently and highly expressed antigen(s) are labeled with different cadmium isotopes and combined in sets of three (x's) to produce LCBs. Cadmium isotopes can be substituted for other metal isotopes and can be tailored around your panel design. If antibodies target the same epitope and their staining signal is weak due to competition for antigen binding, barcoding may perform better with just two isotopes per barcode (e.g., a 7-choose-2 scheme provides 21 barcodes)

comparison of mean PLAYR signal with measurements within bulk samples using quantitative PCR. We have found that perturbation of the targeted transcripts using inhibition or stimulation of an appropriate cell line produces time point samples in which measured RNA quantity closely correlates between both techniques (Fig. 2).

Figure 3 shows a schematic of the antigen and RNA staining procedure by PLAYR. **Methods** Subheading 3.2 provides the steps required to stain surface antigen, label dead cells, and prepare samples prior to PLAYR. If using mass cytometry, LCB can be

Table 2
Oligonucleotide sequences for 27 PLAYR systems

System	3′ Probe1	3′ Probe2	Insert	Detection Probe
1	AAAAAAAAAACTCAGTCGTGACACTCTT	AAAAAAAAAAGACGCTAATATCGTGACC	P-ACGACTGAGTTTGGTCACGAT	Z-ACGACTGAGTTTGGTCACGAT
2	AAAAAAAAAATATCGTCCGGACACTCTT	AAAAAAAAAAGACGCTAATCTTCGAGAC	P-CGGACGATATTTGTCTCGAAG	Z-CGGACGATATTTGTCTCGAAG
3	AAAAAAAAAAGATTCCTCGGACACTCTT	AAAAAAAAAAGACGCTAATCTGCCAATG	P-CGAGGAATCTTTCATTGGCAG	Z-CGAGGAATCTTTCATTGGCAG
4	AAAAAAAAAACTACCTTGGGACACTCTT	AAAAAAAAAAGACGCTAATCAGGCTACT	P-CCAAGGTAGTTTAGTAGCCTG	Z-CCAAGGTAGTTTAGTAGCCTG
5	AAAAAAAAAACTCTTCGAGGACACTCTT	AAAAAAAAAAGACGCTAATCACCAGTTG	P-CTCGAAGAGTTTCAACTGGTG	Z-CTCGAAGAGTTTCAACTGGTG
6	AAAAAAAAAACTTAGCCTGGACACTCTT	AAAAAAAAAAGACGCTAATCCAGACTGT	P-CAGGCTAAGTTTACAGTCTGG	Z-CAGGCTAAGTTTACAGTCTGG
7	AAAAAAAAAACCGCTTATGGACACTCTT	AAAAAAAAAAGACGCTAATCTACATGGC	P-CATAAGCGGTTTGCCATGTAG	Z-CATAAGCGGTTTGCCATGTAG
8	AAAAAAAAAACTCGATCTGGACACTCTT	AAAAAAAAAAGACGCTAATCAACCTGGT	P-CAGATCGAGTTTACCAGGTTG	Z-CAGATCGAGTTTACCAGGTTG
9	AAAAAAAAAATACGTTCCGGACACTCTT	AAAAAAAAAAGACGCTAATCATCCTGAG	P-CGGAACGTATTTCTCAGGATG	Z-CGGAACGTATTTCTCAGGATG
10	AAAAAAAAAACTGCTCATGGACACTCTT	AAAAAAAAAAGACGCTAATCGCAAGTCT	P-CATGAGCAGTTTAGACTTGCG	Z-CATGAGCAGTTTAGACTTGCG
11	AAAAAAAAAATGACTCTCGGACACTCTT	AAAAAAAAAAGACGCTAATCTCGGAATC	P-CGAGAGTCATTTGATTCCGAG	Z-CGAGAGTCATTTGATTCCGAG
12	AAAAAAAAAACTGTCTACGGACACTCTT	AAAAAAAAAAGACGCTAATCACAGTGTC	P-CGTAGACAGTTTGACACTGTG	Z-CGTAGACAGTTTGACACTGTG

(continued)

Table 2
(continued)

System	3′ Probe1	3′ Probe2	Insert	Detection Probe
13	AAAAAAAAAATTC TCCAGGGACACTCTT	AAAAAAAAAAGACGCTAATC TCAATCGG	P-CCTGGAGAATTTCCGATT GAG	Z-CCTGGAGAATTTCCGATT GAG
14	AAAAAAAAAACACTTG TCGGACACTCTT	AAAAAAAAAAGACGCTAA TCAGATGCCT	P-CGACAAGTGTTTAGG CATCTG	Z-CGACAAGTGTTTAGG CATCTG
15	AAAAAAAAAACTTC TGCAGGACACTCTT	AAAAAAAAAAGACGCTAA TCCAGGATCT	P-CTGCAGAAGTTTA GATCCTGG	Z-CTGCAGAAGTTTA GATCCTGG
16	AAAAAAAAAATCTA TCCGGGACACTCTT	AAAAAAAAAAGACGCTAATC TGTAGACC	P-CCGGATAGATTTGGTCTA CAG	Z-CCGGATAGATTTGGTCTA CAG
17	AAAAAAAAAACGCATC TTGGACACTCTT	AAAAAAAAAAGACGCTAATC TGGCACAT	P-CAAGATGCGTTTATGTGC CAG	Z-CAAGATGCGTTTATGTGC CAG
18	AAAAAAAAAATCTCACG TGGACACTCTT	AAAAAAAAAAGACGCTAA TCCTCGAATG	P-CACGTGAGATTTCATTC GAGG	Z-CACGTGAGATTTCATTC GAGG
19	AAAAAAAAAATCGCTAC TGGACACTCTT	AAAAAAAAAAGACGCTAA TCGCCATGAT	P-CAGTAGCGATTTAT CATGGCG	Z-CAGTAGCGATTTAT CATGGCG
20	AAAAAAAAAATACGCTC TGGACACTCTT	AAAAAAAAAAGACGCTAA TCACACTTGG	P-CAGAGCGTATTTC CAAGTGTG	Z-CAGAGCGTATTTC CAAGTGTG
21	AAAAAAAAAACGTC TTACGGACACTCTT	AAAAAAAAAAGACGCTAA TCATGCGACT	P-CGTAAGACGTTTAGTCG CATG	Z-CGTAAGACGTTTAGTCG CATG
22	AAAAAAAAAACCATTCG TGGACACTCTT	AAAAAAAAAAGACGCTAA TCATCAGCGT	P-CACGAATGGTTTACGCT GATG	Z-CACGAATGGTTTACGCT GATG
23	AAAAAAAAAACCTAG TTCGGACACTCTT	AAAAAAAAAAGACGCTAATC TACCTAGG	P-CGAACTAGGTTTCCTAGG TAG	Z-CGAACTAGGTTTCCTAGG TAG
24	AAAAAAAAAACTCCGA TTGGACACTCTT	AAAAAAAAAAGACGCTAA TCAACGCGTT	P-CAATCGGAGTT TAACGCGTTG	Z-CAATCGGAGTT TAACGCGTTG

25	AAAAAAAAAATTCGCAC TGGACACTCTT	AAAAAAAAAAGACGCTAA TCAATTCCGG	P-CAGTGCGAATTTCCG GAATTG	Z-CAGTGCGAATTTCCG GAATTG
26	AAAAAAAAAATCC TTCAGGGACACTCTT	AAAAAAAAAAGACGCTAA TCCGCTAAGT	P-CCTGAAGGATTTACT TAGCGG	Z-CCTGAAGGATTTACT TAGCGG
27	AAAAAAAAAACGTTAC TCGGACACTCTT	AAAAAAAAAAGACGCTAATC TTAAGCGC	P-CGAGTAACGTTTGCGCT TAAG	Z-CGAGTAACGTTTGCGCT TAAG
Universal backbone: P-ATTAGCGTCCAGTGAATGCGAGTCCGTCTAGGAGAGTAGTACAGCAGCCGTCAAGAGTGTC				

P, 5′-phosphorylated; Z, 5′-C6-thiol modification for mass cytometry or fluorophore modification for flow cytometry

Table 3
List of 5′ PLAYR probe sequences that target constituently expressed RNA in human cells

Gene	Probe pair	5′ Probe1	5′ Probe2
GAPDH	1	TGATGGCAACAATATCCACTTT	GTTAAAAGCAGCCCTGGTGA
	2	TGGAAGATGGTGATGGGATT	ATTGATGACAAGCTTCCCGT
	3	TGGACTCCACGACGTACTCA	CATCGCCCCACTTGATTTT
	4	CCTGCTTCACCACCTTCTTG	TCATCATATTTGGCAGGTTTTT
ACTB	1	TGCCAATGGTGATGACCTG	GTCAGGCAGCTCGTAGCTCT
	2	TGTTGGCGTACAGGTCTTTG	ATGTCCACGTCACACTTCATG
	3	CGTCATACTCCTGCTTGCTG	ATCTGCTGGAAGGTGGACAG
	4	TGTTTTCTGCGCAAGTTAGGT	TCAAGAAAGGGTGTAACGCA
RPL27	1	GTTTCATGAACTTGCCCA TTTCGG	GGCCCTACCAGCAAAAAGGAAAG
	2	GCGATCTGAGGTGCCATCATCAA T	TCTTCACGATGACAGCTTTGCG TC
	3	TCTCTTGGCGATCTTCTTC TTGCC	GCTGTCACTTTGCGGGGGTAG
	4	TCTTCAAACTTGACCTTGGCC TCC	GCCTTGCGTTTAAGAGCAGGATC T
	5	AGCATCTAAAACCGCAGTTTC TGG	ACCACTTGTTCTTGCCTGTCTTG T

Table shows predesigned 5′ sequences for PLAYR probe pairs targeting the indicated "housekeeping" transcripts. To complete the PLAYR probe, the 3′ Probe1 or 2 sequence from the PLAYR systems table (Table 2; any system may be chosen) needs to be added to their 3′ end

performed as the first procedure, allowing samples to be pooled into one tube prior to further downstream processing. All incubations for antigen staining are performed with live cells which are placed on ice to avoid any potential changes in transcript expression caused by antibody binding. To reduce loss in surface antibody staining during PLAYR we perform an additional amine-to-amine crosslinking step also on ice. LCB and phenotyping antibody staining, as well as live/dead cell discrimination, are optional and can be individually excluded without affecting one another or the RNA detection by PLAYR. Once stained, cells are treated with paraformaldehyde and permeabilized in methanol, after which there is an opportunity to pause the protocol by storing the sample(s) at −80 °C. Staining the RNA using PLAYR (**Methods** Subheading 3.3) takes approximately 10 h to complete for mass cytometry and 8 h for flow cytometry, with frequent incubation steps. If required, internal antigen staining can be performed in conjunction with the

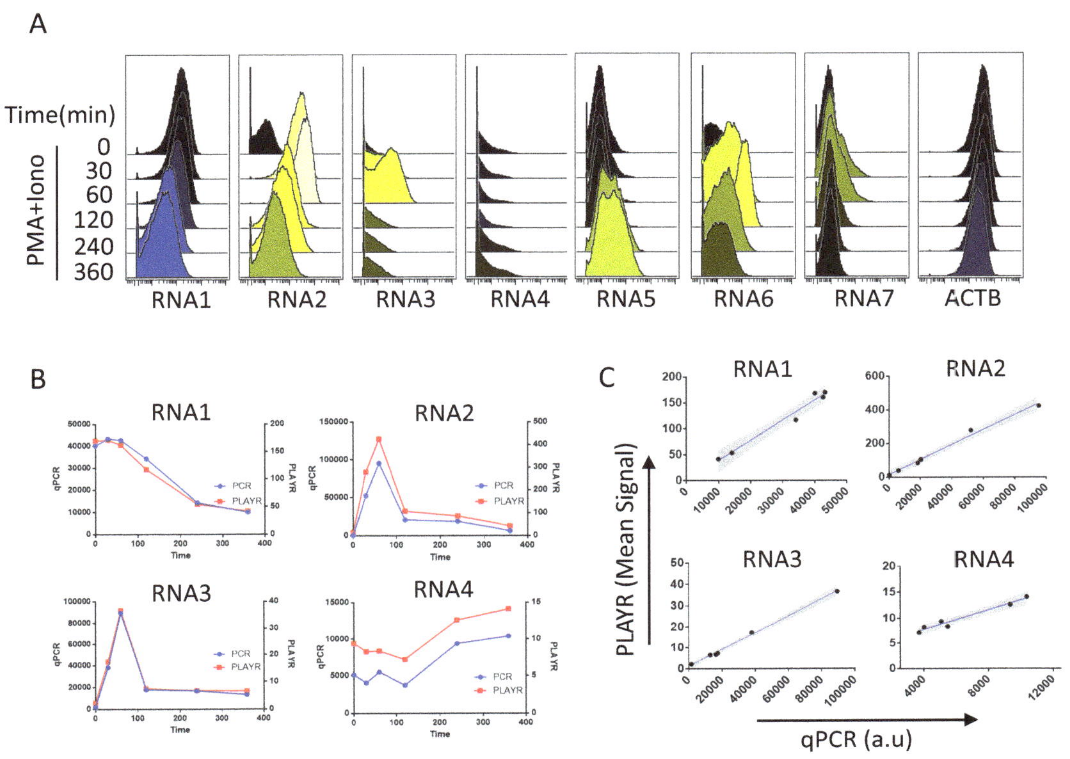

Fig. 2 Validation of PLAYR probe specificity. B-cell line MAVER1 treated with 100 nM phorbol 12-myristate 13-acetate (PMA) + 1 μM ionomycin (PMA + Iono) for up to 6 h was examined by PLAYR (mass cytometry) and qPCR for expression changes in eight RNAs (including ACTB). (**a**) Overlayed histograms showing single-cell expression for each of the eight transcripts at each timepoint. (**b**) Comparison of mean PLAYR signal (right-hand y axis) with quantification by qPCR (left-hand y axis) for 4 RNAs shown in (**a**) over the time course with PMA + Iono. (**c**) Linear regression of data shown in (**b**) showing concordance of relative signal for PLAYR (y axis) and qPCR (x axis) at each timepoint; gray shading indicates 95% confidence intervals

last step of PLAYR where DDP is added. After RNA and internal antigen staining, cells should be run immediately for analysis by flow cytometry. In contrast, cells can be left in fixative and analyzed up to 48 h later using mass cytometry. Instructions on how to process samples for analysis by mass cytometry are described in **Methods** Subheading 3.4 of the protocol. The protocol has the potential to be adapted for non-suspension cells for quantification of RNA in tissues or adherent cells using microscopy.

2 Materials

All buffers are made and stored at room temperature (RT) unless stated otherwise. Filter tips should be used for all mass cytometry and PLAYR methods. All reagents should be RNase and DNase

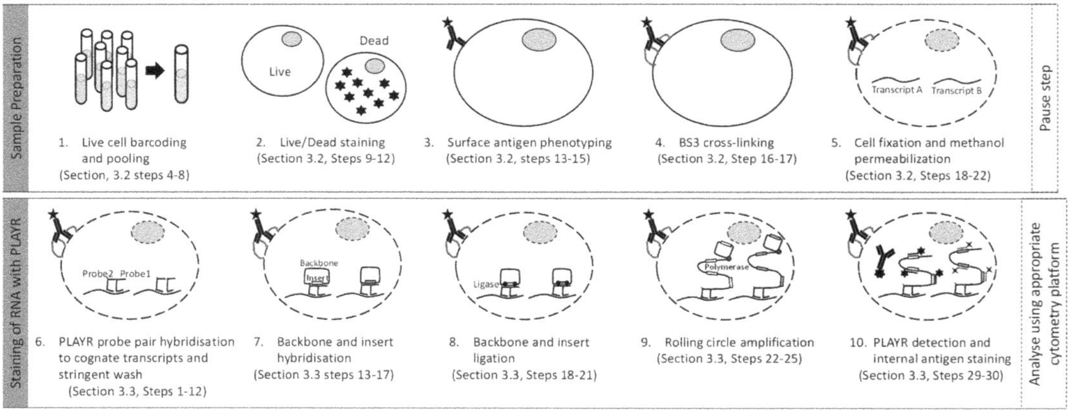

Fig. 3 Schematic of PLAYR staining procedure. Schematic overview of the PLAYR assay, with the top panel showing the stages (1–5) for sample preparation and the bottom panel illustrating the stages (6–10) for RNA staining with PLAYR. Stages 1–4 are all optional and, if not required, can be excluded without affecting the detection of RNA with PLAYR. For PLAYR-CYTOF, live cell barcoding can be performed first (stage 1), allowing multiple samples to be pooled into one tube. Dead cells can then be stained at RT (stage 2) followed by surface antigen staining on ice (stage 3). Covalent crosslinking of surface antibodies using BS3 reagent is then performed (stage 4) on ice to help maintain antigen binding during the remaining stages of PLAYR. Cells are then fixed in PFA and permeabilized with methanol in stage 5. Once in methanol, cells can be stored at −80 °C allowing the assay to be paused. Stages 6–10 illustrate the sequential steps of oligonucleotide binding and amplification used by PLAYR. Internal antibody staining can be included in the final stage alongside incubation with DDPs

free. If performing PLAYR-CyTOF, then extra care must be taken to ensure the reagents are metal free, particularly those used in the final wash steps of the assay; Fluidigm supplies ddH_2O, phosphate buffer saline (PBS), and cell staining buffer (CSB) that are all guaranteed to be metal free.

2.1 Labeling DPPs with Metal Isotopes for Mass Cytometry

1. Microcentrifuge.
2. Water bath.
3. Ice-cold 100% ethanol.
4. Ice-cold 75% ethanol.
5. 3 M Sodium acetate: Dissolve in ddH_2O and adjust the pH to 5.2 with glacial acetic acid.
6. 500 mM Tris(2-carboxyethyl)phosphine hydrochloride (TCEP) solution.
7. Maxpar X8 Antibody Labeling Kit (Fluidigm). The choice of metal for each DDP depends upon your panel design.
8. NanoDrop spectrophotometer.
9. Centrifugal filter with 30 kDa molecular weight cut off (MWCO).
10. Centrifugal filter with 3 kDa MWCO.

11. 250 μM DDP carrying a 5′-thiol-modifier C6 S-S: Purified by HPLC and dissolved in ddH_2O. Store at −20 °C. (*See* Table 2 for list of sequences.)

2.2 General Materials for PLAYR

1. Refrigerated microcentrifuge (swing rotor is advised, *see* **Note 1**).
2. ddH_2O.
3. PBS (RT and ice cold).
4. CSB: 0.5% bovine serum albumin dissolved in PBS. Store at 4 °C. This reagent is required if performing PLAYR-CyTOF or surface antibody staining.
5. Saline Sodium Cirtrate (SSC, 20×).

2.3 LCB for PLAYR-CyTOF (Optional)

1. Metal-conjugated antibodies for LCB: Dilute antibodies in CSB to make 100 μL barcoding solution per 3×10^6 cells stained (*see* **Note 3**). Antibody solutions can be made 2 h before use and stored at 4 °C. See Table 1 for an example of preparing an LCB scheme for PBMC using cadmium-conjugated anti-CD45 antibodies.

2.4 Staining to Identify Dead Cells (Optional)

1. Dead cell exclusion reagent: For mass cytometry we use Cell-ID cisplatin (Fluidigm, cat. no. 201064). For flow cytometry we use Live/Dead Fixable Dead Cell Stain Kits diluted 1:1000 in RT PBS (Invitrogen). Reagents should be diluted 1:1000 immediately before use in RT PBS.

2.5 Surface Antibody Staining (Optional)

1. Metal- (for mass cytometry [19]; Fluidigm) or fluorophore-conjugated (for flow cytometry) antibodies for surface antigen phenotyping (*see* **Note 2**): Dilute antibodies in CSB to make 100 μL of antigen phenotyping solution per 3×10^6 cells stained (*see* **Note 3**). Antibody cocktails can be made up to 2 h before use and stored at 4 °C.
2. 5 mM Bis(sulfosuccinimidyl) suberate (BS3 crosslinker): Dissolve BS3 crosslinker in ice-cold PBS (*see* **Note 4**).

2.6 Cell Fixation and Permeabilization

1. Ice-cold 1.6% Paraformaldehyde (PFA): Dilute 16% PFA using ice-cold PBS and place on ice. 1.6% PFA can be made up to 1 day before use.
2. Ice-cold methanol.

2.7 PLAYR and Internal Antibody Staining

1. Thermal mixer (e.g., Eppendorf ThermoMixer C, cat. no. 5382000031).
2. 100 μM PLAYR oligonucleotides: PLAYR probes should be dissolved in ddH_2O and aliquoted (*see* **Note 5**). Purification by

standard desalting is sufficient for PLAYR probes, while 5′--phosporylated oligonucleotides (i.e., backbone and inserts) should be purified by HPLC. See Table 2 for list of sequences (*see* also Table 3). Oligonucleotides should be stored at −20 °C for up to 3 years.

3. PLAYR dilution buffer (PBST): 0.1% Tween diluted in PBS. Make fresh at the start of the PLAYR assay (Subheading 3.3, **Step 1**) and place on ice.
4. PLAYR wash buffer (PBSTR): 4 U/mL RNasin Plus (Promega, cat. no. N2611/N2615) (*see* **Note 6**) diluted in PBST. Make fresh at the start of the PLAYR assay (Subheading 3.3, **Step 1**) and place on ice.
5. PLAYR wash buffer with extra RNasin (PBSTR+): 40 U/mL RNasin Plus (Promega, cat. no. N2611/N2615) (*see* **Note 6**) diluted in ice-cold PBST. Make fresh at the start of the PLAYR assay (Subheading 3.3, **Step 1**) and place on ice.
6. Preliminary PLAYR probe hybridization buffer (*see* **Note 7**): 1× SSC, 1% Tween20 (*see* **Note 8**), 2.5% Polyvinylsulfonic acid, and 100 μg/mL Salmon sperm DNA. Reagents are diluted in the appropriate amount of ddH_2O (*see* **Note 7**). Make fresh at the start of the PLAYR assay (Subheading 3.3, **Step 3**) and leave at RT.
7. Final PLAYR probe hybridization buffer (*see* **Note 7**): Preliminary PLAYR probe hybridization buffer plus 100 nM of each PLAYR probe which has been preheated (*see* **Note 5,** preheated in Subheading 3.3, **Step 5**), 20 mM Ribonucleoside vanadyl complexes (*see* **Note 9**, preheated in Subheading 3.3, **Step 2**), and 40 U/mL RNasin Plus (Promega, cat. no. N2611/N2615) (*see* **Note 6**). Make this immediately before adding to the cells in Subheading 3.3, **Step 6**.
8. PLAYR post-hybridization wash buffer: 4× SSC (*see* **Note 10**) and 40 U/mL RNasin diluted in ice-cold PBSTR. Make fresh up to 20 min before use in Subheading 3.3, **Step 9**.
9. PLAYR backbone/insert hybridization mix: 1× SSC, 40 U/mL RNasin, 100 nM backbone oligonucleotide, and 100 nM of each insert oligonucleotide. Dilute in ice-cold PBSTR. Make fresh up to 20 min before use in Subheading 3.3, **Step 12**.
10. PLAYR T4 buffer wash: 1× T4 ligase buffer (as supplied by the manufacturer) and 40 U/mL RNasin. Make fresh up to 20 min before use in Subheading 3.3, **Step 16**.
11. PLAYR T4 ligation mix: 1× T4 ligase buffer (as supplied by the manufacturer), 5 U/mL T4 ligase and 40 U/mL RNasin. Dilute in ddH_2O. Make fresh up to 20 min before use in Subheading 3.3, **Step 16**.

12. PLAYR Phi29 buffer wash: 1× Phi29 buffer (as supplied by the manufacturer) diluted in ddH_2O. Make fresh up to 20 min before use in Subheading 3.3, **Step 20**.
13. PLAYR Phi29 amplification mix: 1× Phi29 buffer (as supplied by the manufacturer), 5 U/mL Phi29 DNA polymerase, and 800 μM dNTP. Dilute in RNase-free ddH_2O. Make fresh up to 20 min before use in Subheading 3.3, **Step 20**.
14. PLAYR mass cytometry detection probe buffer: 10 nM of each metal-labeled DDP diluted in cell staining buffer. Use this buffer if performing analysis by mass cytometry. If internal antigens are also being analyzed, an appropriate concentration of metal-labeled antibodies should be added to this buffer. Make fresh up to 2 h before use in Subheading 3.3, **Step 24**.
15. PLAYR flow cytometry detection probe buffer: 1 nM of each fluorescent-labeled DDP diluted in CSB. Use this buffer if performing analysis by flow cytometry. If internal antigens are also being analyzed, an appropriate concentration of fluorescent-labeled antibodies should be added to this buffer. Make fresh up to 2 h before use in Subheading 3.3, **Step 24**.

2.8 Additional Reagents Specific to Mass Cytometry

1. 62.5 nM Cell-ID Intercalator-Ir buffer: Dilute Cell-ID Intercalator-Ir (Fluidigm, cat. no. 201192A) in PBS containing 1.6% PFA.
2. EQ Four Element Calibration Beads (Fluidigm, cat. no. 201078).
3. Cell acquisition solution (CAS, Fluidigm, cat no. 201240).
4. 5 mL Round-Bottom Polystyrene Test Tubes with 35 μm Cell Strainer Cap.

2.9 Software

1. Fluidigm CyTOF system control software for mass cytometry (www.fluidigm.com/software).
2. NxT software (Thermofisher; cat. no. A25554) for the Attune flow cytometer (or equivalent).
3. Software appropriate for analysis of multiparametric mass cytometry data files (e.g., Cytobank (https://www.cytobank.org/), Cytosplore (https://www.cytosplore.org/), etc.).
4. R software (https://www.r-project.org/) for PLAYR probe design.
5. PLAYRDesign software to design PLAYR probes for your RNA of interest (detailed instructions for installation can be found at https://github.com/nolanlab/PLAYRDesign).

3 Methods

3.1 Labeling DDPs with Metal Isotopes for Mass Cytometry

1. Calculate or make a note of the starting number of nmol for each detection probe as this will be required during calculations in **step 18**.
2. Add 500 mM TCEP to the 250 μM DDP to a final concentration of 50 mM (1 in 10) and incubate at RT for 30 min.
3. If required, raise the volume to a minimum of 200 μL using ddH_2O and then add sodium acetate to a final concentration of 300 mM. Mix by vortexing.
4. Add three volumes of ice-cold 100% ethanol and mix by vortex.
5. Place the tube at −20 °C overnight to allow precipitation. Alternatively, precipitation and longer-term storage can be performed at −80 °C for up to 6 months.
6. From the Maxpar X8 Antibody Labeling Kit, centrifuge the X8 polymer tube 4000rcf/30 s/RT) to ensure the reagent is at the bottom.
7. Dissolve the polymer in 95 μL of L-Buffer by pipetting.
8. Add 5 μL of lanthanide metal solution and mix by pipetting.
9. Incubate the polymer-metal mix at 37 °C for 30 min.
10. While the polymer-metal mix is incubating (**step 9**), centrifuge the precipitated DDP from **step 5** at 14,000rcf/4 °C/30 min.
11. Add 200 μL of L-Buffer to a 3 kDa MWCO filter.
12. After the incubation in **step 9**, add the metal-loaded polymer to the filter containing the 200 μL L-Buffer.
13. Centrifuge metal-conjugated polymer at 12,000rcf/RT/ 30 min (*see* **Note 11**).
14. Once the centrifuge step is completed from **step 10**, carefully remove supernatant from DDP pellet. Resuspend in 400 μL ice-cold 75% ethanol and centrifuge at 14,000rcf/4 °C/ 10 min.
15. Retrieve 3 kDa filter from centrifuge (from **step 13**) and discard flow through. Add 300 μL of C-Buffer to the filter. Centrifuge at 12,000rcf/RT/30 min; leave at RT while completing detection probe purification.
16. Repeat **step 14** to wash the DDP pellet a second time.
17. After the final wash of the DDP in **step 16**, remove as much ethanol as possible (a brief centrifuge step (14,000rcf/4 °C/ 30 s) may help to pool residual ethanol for removal). Air-dry the pellet (the tube can be placed at 40 °C on a heat block to decrease drying time, *see* **Note 12**).

18. Resuspend the dried pellet of reduced DDP in C buffer (from the Maxpar X8 Antibody Labelling Kit) to a concentration of 1 nmol/25 μL, assuming a yield of 80% after reduction. For example, if there was 10 nmol of DDP to start, then you would assume a yield of 8 nmol after the reduction/purification steps and therefore resuspend the pellet in 200 μL of C buffer.
19. Measure the reduced DDP concentration from **step 18** by absorption at 260 nM on a NanoDrop spectrophotometer using C buffer as a blank.
20. Using the concentration from **step 19**, calculate the volume required for 2 nmol of DDP (MW = 6789 g/mol) and transfer this volume to a fresh Eppendorf tube.
21. Transfer the entire metal-loaded polymer from the 3 kDa filter to the Eppendorf tube containing the 2 nmol DDP.
22. Make the volume up to 100 μL using C-Buffer and mix by pipetting.
23. Incubate at RT for 2 h.
24. Add 500 mM TCEP 1 in 100 to a final concentration of 5 mM, immediately mix by vortexing, and incubate at RT for 30 min.
25. Add 400 μL of ddH_2O to a 30 kDa MWCO filter.
26. Transfer the metal-polymer-DDP mixture to the 30 kDa MWCO filter containing the ddH_2O and centrifuge at 14,000rcf/RT/12 min.
27. Discard the flow-through, add 500 μL of ddH_2O to the filter, and centrifuge at 14,000rcf/RT/12 min.
28. Repeat **step 27**.
29. Invert the filter into a new collection tube and centrifuge the filter and tube at 1000rcf/RT/2 min.
30. Wash the filter with 250 μL of ddH_2O and elute into the same collection tube by inverting the filter and centrifuging the filter and tube at 1000rcf/RT/2 min.
31. Measure the oligonucleotide concentration by NanoDrop spectrophotometer using ddH_2O as a blank.
32. Adjust the volume of ddH_2O to obtain a final concentration of 1 μM metal-labeled DDP (assuming 6789 ng/mL = 1 μM).
33. Aliquot and store the metal-labeled DDP at −20 °C for up to 3 years.

3.2 Surface Antibody Staining, Dead Cell Exclusion, and Preparing Cells for PLAYR

LCB for PLAYR-CyTOF, staining to identify dead cells, and surface antibody staining are all independent staining procedures and, if not required, each section can be excluded without affecting one another or the detection of RNA by PLAYR. To minimize changes in transcript expression during surface antibody staining, cells should be kept cold (between 1 °C and 4 °C) for all steps unless stated otherwise. All incubation steps for flow cytometry during and post binding of fluorophore-conjugated reagents should be performed in the dark to reduce quenching.

1. Aliquot 3×10^6 cells into an Eppendorf tube on ice for each sample (*see* **Notes 13** and **14**).
2. Pellet cells by centrifugation at 500rcf/4 °C/3 min. (*See* **Notes 15** and **16**.)
3. Wash in ice-cold PBS by centrifugation at 500rcf/4 °C/3 min.

LCB for PLAYR-CyTOF (Optional)

4. Resuspend each sample in its appropriate LCB antibody solution (3×10^6 cells per 100 μL) and incubate on ice for 30 min. Resuspend cells by agitation after 15 min.
5. Pellet cells by centrifugation at 500rcf/4 °C/3 min.
6. Wash pellet twice in 200 μL ice-cold CSB.
7. Resuspend in 200 μL ice-cold PBS and pool barcoded samples together into one tube.
8. Centrifuge at 500rcf/4 °C/5 min and remove supernatant.

Staining to Identify Dead Cells (Optional)

9. Resuspend cells in 1 mL dead cell exclusion reagent per 10^7 cells.
10. Incubate at RT for 5 min.
11. Add 4× volume of ice-cold CSB to quench unreacted dead cell exclusion dye and place tubes on ice.
12. Pellet cells by centrifugation at 500rcf/4 °C/5 min and remove supernatant.

Surface Antibody Staining (Optional)

13. Resuspend cells in surface antibody cocktail and incubate on ice for 30 min. Resuspend cells by agitation after 15 min.
14. Wash cells twice in ice-cold CSB.
15. Wash cells in ice-cold PBS.
16. Resuspend cells in 5 mM BS3 crosslinker (1 mL per 10^7 cells) and incubate on ice for 30 min (*see* **Note 4**). Resuspend cells by agitation after 15 min.
17. Pellet cells by centrifugation at 500rcf/4 °C/3 min and remove supernatant.

Cell Fixation and Permeabilization

18. Resuspend cells in ice-cold 1.6% PFA (1 mL per 3×10^6 cells) and incubate on ice for 10 min.

19. Centrifuge cells at 800rcf/4 °C/3 min and remove supernatant.
20. Flick pellet to resuspend cells in residual supernatant.
21. Add ice-cold methanol dropwise while vortexing the sample (*see* **Note 17**).
22. Incubate on ice for 10 min. Alternatively, cells suspended in methanol can be placed at −80 °C and the protocol can be paused (*see* **Note 18**).

3.3 PLAYR and Internal Antibody Staining

All centrifuge steps are performed at 800rcf/4 °C/3 min unless stated otherwise.

1. Prepare PBST, PBSTR, and PBSTR+ buffers and place on ice.
2. Heat RVC at 65 °C for 15 min with vigorous agitation to dissolve any precipitation and then place at RT (do not place on ice as it will cause the RVC to reprecipitate).
3. While RVC is heating, prepare the preliminary PLAYR probe hybridization buffer.
4. Once the RVC has finished heating and been placed at RT, pellet cells in methanol (*see* **Note 19)** and wash with 200 μL of ice-cold PBSTR+ (*see* **Notes 20** and **21**).
5. While washing cells in the previous step, heat PLAYR probe mixture at 90 °C for 5 min and then place immediately on ice (*see* **Note 22**).
6. Quickly add remaining reagents (preheated RVC, preheated PLAYR probes, and RNasin Plus) to make the final PLAYR hybridization buffer and immediately use to resuspend cells at a maximum of 3×10^7 cells/mL (minimum of 100 μL).
7. Incubate cells for 1 h at 40 °C (*see* **Note 23**) with vigorous agitation (we use 1400 rpm).
8. Wash cells three times with 200 μL of ice-cold PBSTR.
9. While washing the cells in **step 8** prepare PLAYR post-hybridization wash buffer.
10. Resuspend cells in PLAYR post-hybridization wash buffer at a maximum of 3×10^7 cells/mL (minimum of 100 μL) and incubate at 40 °C (*see* **Note 23**) for 20 min under vigorous agitation (we use 1400 rpm).
11. During this incubation prepare the PLAYR backbone/insert hybridization mix.
12. Wash cells two times in ice-cold PBSTR.
13. Resuspend the cells in PLAYR backbone/insert hybridization mix at 3×10^7 cells/mL (minimum of 100 μL).

14. Incubate cells for 30 min at 37 °C with agitation (we use 1100 rpm).
15. While incubating prepare the 1× PLAYR T4 wash buffer and PLAYR T4 ligation mix.
16. Wash cells two times in 200 μL ice-cold PBSTR.
17. After two washes in PBSTR, wash cells in 100 uL of 1× PLAYR T4 wash buffer.
18. Resuspend the cells in PLAYR ligation mix at 6×10^7 cells/mL (minimum of 50 μL).
19. Incubate cells for 30 min at RT (21 °C) with agitation (we use 1100 rpm).
20. While incubating prepare the Phi29 wash buffer and PLAYR Phi29 amplification mix.
21. Wash the cells in Phi29 wash buffer.
22. Resuspend the cells in PLAYR Phi29 amplification mix at 6×10^7 cells/mL (minimum of 50 μL).
23. Incubate cells for 2 h (flow cytometry) or 4 h (mass cytometry) at 30 °C with agitation (we use 1100 rpm). See **Note 24**.
24. Prepare either the PLAYR mass- or flow-cytometry detection probe buffer.
25. Wash cells once in 200 μL PBSTR (flow cytometry) or CSB (mass cytometry).
26. Resuspend cells in the appropriate PLAYR detection probe buffer at 3×10^7 cells/mL (minimum of 100 μL).
27. Incubate cells for 30 min at 37 °C with agitation (we use 1100 rpm).
28. Wash cells twice in PBSTR (for flow cytometry) or CSB (for mass cytometry).
29. For flow cytometry, resuspend cells in ice-cold PBS and analyze immediately using laboratory-specific protocols. For mass cytometry, resuspend cells in Cell-ID Intercalator-Ir buffer and incubate for 1 h at room temperature or for up to 2 weeks at 4 °C (*see* **Note 25**).

3.4 Sample Preparation for Mass Cytometry

1. Wash cells once in 1 mL CSB (800rcf/4 °C/3 min).
2. Wash cells once in 1 mL CAS (800rcf/4 °C/3 min).
3. Resuspend in 1 mL CAS.
4. Filter through a 35 μM filter cap into a 5 mL test tube.
5. Count cells and adjust concentration to 5×10^5 cells/mL using CAS (*see* **Note 26**).
6. Add 1/10 volume of EQ Four Element Calibration Beads and mix by vortexing (*see* **Note 27**).

7. Collect data using the mass cytometer. Details on instrument and panel-template setup are outside the scope of this protocol but can be found in the user guides on the Fluidigm website.
8. Analyze data using software such as cytobank, flowjo, or relevant R packages.

4 Notes

1. Due to the large number of centrifuge steps in the protocol, every effort must be made to minimize cell loss. Swing rotor centrifuges are highly recommended alongside the use of small wash volumes (200 μL) and Polypropylene V-bottom 200 μL PCR strips, or alternatively 1.5 mL Eppendorf tubes. Always remove supernatant by aspiration (i.e., not by flicking the liquid out) and use small pipette tips (200 μL or lower) to increase accuracy. If the supernatant volume is high (i.e., >2 mL), it may be advised to remove down to ~300 μL, resuspend the pellet again, and centrifuge a second time.
2. Metal- and fluorophore-conjugated antibodies used to stain surface antigen for measurement by flow or mass cytometry need to withstand conditions used during the remaining steps of the protocol (i.e., PFA fixation, methanol treatment, and PLAYR incubations). For flow cytometry, this generally means the use of smaller fluorophores (e.g., FITC, Alexa Fluors) which are not protein based and that are much more likely to remain functional during the procedure. For PLAYR-CyTOF, Bismuth-209 conjugated with X8 polymers is the only metal/polymer conjugate found so far that cannot withstand the PLAYR conditions. Metals and fluorophores employed at the end of the protocol (i.e., conjugated to internal antigens or DDPs) are not subjected to harsh conditions, and therefore their choice is not limited at this later step.
3. Antibodies should be serially diluted to confirm specific and efficient staining of target cells after PLAYR. Generally, final concentrations of staining antibodies fall in the range of 10–1.25 μg/mL (most often 0.5 μg/mL is used). For LCB, the signal produced by positively stained cells should be clearly distinct from the signal generated by cells that have not been stained. If the LCB antibodies target the same epitope, their dilution should be tested in the barcoding combinations (as using multiple antibodies that compete for binding to the same epitope will reduce signal for each barcoding metal).
4. BS3 Crosslinker hydrolyzes in water. Allow the vial containing BS3 to reach RT before opening to avoid condensation. BS3

crosslinker powder should be weighed and dissolved in ice-cold PBS immediately before addition to the cells.

5. To increase efficiency of PLAYR probe hybridization buffer preparation we recommend that the 100 μM PLAYR probes that target the same transcript are mixed in equal volumes, and aliquots are made, so that the volume of the aliquots (in μL) is equal to the number of probes being used for that transcript (e.g., if four probe pairs are being used to target a transcript then 8 μL aliquots would be made as there are eight PLAYR probes). Each aliquot can then be frozen and will be sufficient for 1 mL of PLAYR probe hybridization buffer. If the same set panel of transcripts will be analyzed repeatedly, then to speed the process up even further, each transcript-specific aliquot made above can be pooled in advance of the PLAYR assay and stored frozen (e.g., 10 transcripts are being analyzed, each being targeted by 8 PLAYR probes, would produce an 80 μL aliquot of PLAYR probe mixture that can be heated and added to the final PLAYR probe hybridization buffer).
6. We have tested other RNasin brands and found that RNasin Plus (Promega) is the best at maintaining RNA integrity during the PLAYR assay. Other RNasins may give suboptimal results.
7. To standardize the process (in terms of PLAYR probe and RVC aliquot sizes) we recommend making up the PLAYR probe hybridization buffer in 1 mL volumes (where 1 mL is enough for 3×10^7 cells). The PLAYR probe hybridization buffer is made up in two stages (termed the preliminary and final). The preliminary buffer is made shortly before use (in Subheading 3.3, **Step 3**) but lacks the preheated PLAYR probes, preheated RVC and RNasin plus that will need to be added immediately before addition to the cells (Subheading 3.3., **Step 6**). The volume of ddH_2O required in the preliminary buffer is dependent on the volume of probes added to the final buffer to give a total volume of 1 mL.
8. 1% Tween20 can be replaced with lower concentrations of the detergent (we have reduced this to 0.2% Tween20 without noticing any difference in signal for our PLAYR panel). Specificity for your specific PLAYR probes should be confirmed while using these milder conditions.
9. RVC precipitates at low temperatures and requires heating at 65 °C for 15 min to dissolve and fully activate the reagent. The RVC will arrive frozen from the supplier and should be defrosted and heated in a 65 °C water bath with frequent agitation to allow it to dissolve. Once dissolved, aliquots can be generated at RT and then frozen at −80 °C. For a 200 mM

RVC solution, a 100 μL aliquot will be sufficient to make 1 mL of PLAYR probe hybridization buffer which is enough for 1×10^7 cells or up to 10 samples (*see* **Note 7** on making the PLAYR probe hybridization buffer).

10. 4× SCC can be replaced with 2× SCC concentration to help maintain surface antibody binding. Specificity for your specific PLAYR probes should be confirmed while using these milder conditions.
11. You should aim for a volume after centrifugation that is lower than 20 μL. If it is higher, then centrifuge again for 5–10 min.
12. Failure to fully dry the pellet and remove all ethanol can affect downstream applications. The pellet is dry when it turns from opaque/white to bright white in color.
13. We encourage a minimum of 3×10^6 cells per sample because of cell loss throughout the assay. This yields approximately (depending on pipetting efficiency) 1×10^6 cells at the final wash step to analyze by mass/flow cytometry. Lower cell numbers per sample can be used for mass cytometry if the experimental plan includes LCB and pooling.
14. Flow cytometry experiments will require a negative control sample to account for autofluorescence and background staining of the DDP. This could be a sample that is assayed without probe 1 or 2 (i.e., only one of the probe pair is present), or alternatively the backbone sequence is excluded from the backbone/insert hybridization mix.
15. Due to the considerable number of centrifuge steps during antibody and RNA staining we highly recommend that cells are placed in small tubes and pelleted in swing rotor centrifuges using small volumes (200 μL or less) where possible to avoid excessive cell loss. Small volumes (<20 μL) of supernatant may be left during repeat wash steps to decrease cell loss.
16. Centrifugation details given are for primary PBMC cells and can be modified for your cell of interest if required.
17. To avoid cell clumping, cells should be in a uniform suspension while methanol is added.
18. Fixed cells can remain in methanol at −80 °C for several months without noticeable depreciation in subsequent antibody or RNA signal.
19. Methanol has increased density when at −80 °C, which may inhibit the pelleting of cells. Allow the methanol to warm up to ice temperature for 5 min before centrifugation.

20. Once cells are removed from methanol RNA degradation can set in. Speed in processing during the PBSTR+ wash step and fast transfer into the PLAYR probe hybridization buffer is essential to maintain RNA integrity.
21. Due to the small volumes used, ensure that all wash buffer is removed from the pellet before the addition of PLAYR hybridization buffer to the cells in Subheading 3.3 **Step 6**.
22. Cool on ice to avoid denaturing RNasin in the PLAYR hybridization buffer.
23. Incubation temperatures of PLAYR hybridization buffer and post hybridization wash can be reduced to 37 °C to help maintain surface antibody binding. Specificity for your specific PLAYR probes should be confirmed while using these milder conditions.
24. The RCA times using Phi29 can be made longer to increase the signal from the assay. However, we have found that longer RCA times (past the 2 h and 4 h suggested) do not increase sensitivity of the assay (i.e., negative cells remain negative). As mentioned in the introduction, the best way to increase sensitivity is by increasing the number of probe pairs used to target the RNA.
25. PLAYR-CyTOF signals will slowly depreciate the longer you leave the samples at 4 °C. We recommend running the samples within 48 h of finishing the assay to achieve maximum signal.
26. Cell concentrations may be adjusted depending upon how readily they clog the mass cytometer. Larger or more sticky cells may require running at lower concentrations.
27. Vigorously shake the EQ Four Element Calibration Bead bottle for 30 s to make sure the contents are evenly resuspended before adding to the sample.

References

1. Bauman JG, Bayer JA, van Dekken H (1990) Fluorescent in-situ hybridization to detect cellular RNA by flow cytometry and confocal microscopy. J Microsc 157(Pt 1):73–81. https://doi.org/10.1111/j.1365-2818.1990.tb02948.x
2. Porichis F, Hart MG, Griesbeck M, Everett HL, Hassan M, Baxter AE, Lindqvist M, Miller SM, Soghoian DZ, Kavanagh DG, Reynolds S, Norris B, Mordecai SK, Nguyen Q, Lai C, Kaufmann DE (2014) High-throughput detection of miRNAs and gene-specific mRNA at the single-cell level by flow cytometry. Nat Commun 5:5641. https://doi.org/10.1038/ncomms6641
3. Lalli E, Gibellini D, Santi S, Facchini A (1992) In situ hybridization in suspension and flow cytometry as a tool for the study of gene expression. Anal Biochem 207(2):298–303. https://doi.org/10.1016/0003-2697(92)90015-y
4. Patterson BK, Till M, Otto P, Goolsby C, Furtado MR, McBride LJ, Wolinsky SM (1993) Detection of HIV-1 DNA and messenger RNA in individual cells by PCR-driven in situ hybridization and flow cytometry. Science 260(5110):976–979. https://doi.org/10.1126/science.8493534
5. Belloc F, Durrieu F (1994) Detection of mRNA species by flow cytometry. Methods

Cell Biol 42(Pt B):59–69. https://doi.org/10.1016/s0091-679x(08)61068-7

6. Borzi RM, Piacentini A, Monaco MC, Lisignoli G, Degrassi A, Cattini L, Santi S, Facchini A (1996) A fluorescent in situ hybridization method in flow cytometry to detect HIV-1 specific RNA. J Immunol Methods 193(2):167–176. https://doi.org/10.1016/0022-1759(96)00070-1
7. Larsson C, Grundberg I, Soderberg O, Nilsson M (2010) In situ detection and genotyping of individual mRNA molecules. Nat Methods 7(5):395–397. https://doi.org/10.1038/nmeth.1448
8. Player AN, Shen LP, Kenny D, Antao VP, Kolberg JA (2001) Single-copy gene detection using branched DNA (bDNA) in situ hybridization. J Histochem Cytochem 49(5):603–612. https://doi.org/10.1177/002215540104900507
9. Frei AP, Bava FA, Zunder ER, Hsieh EW, Chen SY, Nolan GP, Gherardini PF (2016) Highly multiplexed simultaneous detection of RNAs and proteins in single cells. Nat Methods 13(3):269–275. https://doi.org/10.1038/nmeth.3742
10. Duckworth AD, Gherardini PF, Sykorova M, Yasin F, Nolan GP, Slupsky JR, Kalakonda N (2019) Multiplexed profiling of RNA and protein expression signatures in individual cells using flow or mass cytometry. Nat Protoc 14(3):901–920. https://doi.org/10.1038/s41596-018-0120-8
11. Fredriksson S, Gullberg M, Jarvius J, Olsson C, Pietras K, Gustafsdottir SM, Ostman A, Landegren U (2002) Protein detection using proximity-dependent DNA ligation assays. Nat Biotechnol 20(5):473–477. https://doi.org/10.1038/nbt0502-473
12. Soderberg O, Gullberg M, Jarvius M, Ridderstrale K, Leuchowius KJ, Jarvius J, Wester K, Hydbring P, Bahram F, Larsson LG, Landegren U (2006) Direct observation of individual endogenous protein complexes in situ by proximity ligation. Nat Methods 3(12):995–1000. https://doi.org/10.1038/nmeth947
13. Lizardi PM, Huang X, Zhu Z, Bray-Ward P, Thomas DC, Ward DC (1998) Mutation detection and single-molecule counting using isothermal rolling-circle amplification. Nat Genet 19(3):225–232. https://doi.org/10.1038/898
14. Brodin P (2019) The biology of the cell – insights from mass cytometry. FEBS J 286(8):1514–1522. https://doi.org/10.1111/febs.14693
15. Bendall SC, Simonds EF, Qiu P, Amir el AD, Krutzik PO, Finck R, Bruggner RV, Melamed R, Trejo A, Ornatsky OI, Balderas RS, Plevritis SK, Sachs K, Pe'er D, Tanner SD, Nolan GP (2011) Single-cell mass cytometry of differential immune and drug responses across a human hematopoietic continuum. Science 332(6030):687–696. https://doi.org/10.1126/science.1198704
16. Muftuoglu M, Li L, Liang S, Mak D, Lin AJ, Fang J, Burks JK, Chen K, Andreeff M (2021) Extended live-cell barcoding approach for multiplexed mass cytometry. Sci Rep 11(1):12388. https://doi.org/10.1038/s41598-021-91816-w
17. Hartmann FJ, Simonds EF, Bendall SC (2018) A universal live cell barcoding-platform for multiplexed human single cell analysis. Sci Rep 8(1):10770. https://doi.org/10.1038/s41598-018-28791-2
18. Mei HE, Leipold MD, Schulz AR, Chester C, Maecker HT (2015) Barcoding of live human peripheral blood mononuclear cells for multiplexed mass cytometry. J Immunol 194(4):2022–2031. https://doi.org/10.4049/jimmunol.1402661
19. Han G, Spitzer MH, Bendall SC, Fantl WJ, Nolan GP (2018) Metal-isotope-tagged monoclonal antibodies for high-dimensional mass cytometry. Nat Protoc 13(10):2121–2148. https://doi.org/10.1038/s41596-018-0016-7

Chapter 11

Array-Based Comparative Genomic Hybridization for the Detection of Copy Number Alterations in Single Cells

Giancarlo Feliciello, Zbigniew Tadeusz Czyz, and Bernhard M. Polzer

Abstract

Comprehensive genome-wide analyses of single cells represent an important tool for clinical applications, such as pre-implantation diagnostic and prenatal diagnosis, as well as for cancer research purpose. For the latter, studies of tumor heterogeneity, circulating tumor cells (CTCs), and disseminated cancer cells (DCCs) require the analysis of single-cell genomes. Here we describe a reliable and robust array-based comparative genomic hybridization (aCGH) protocol based on *Ampli I*™ whole genome amplification that allows the detection of copy number alterations (CNAs) in single cancer cells as small as 100 kb.

Key words aCGH, Single cells, PCR labeling, Array hybridization, CNAs

1 Introduction

Array comparative genomic hybridization (aCGH) is a technique that enables genome-wide high-resolution screening of genomic copy number alterations (CNAs) [1, 2]. It allows for comprehensive analysis at multiple genomic loci by detecting DNA copy number gains and losses. Nowadays, aCGH is an essential tool in pre-implantation diagnostics [3, 4] and prenatal diagnosis [5]. These diagnostic applications require the analysis of copy number changes at the single-cell level. Similarly, in cancer research, single-cell technologies are more and more demanded for the characterization of circulating tumor cells (CTCs) isolated from blood and disseminated cancer cells (DCCs) found in a secondary distant site (e.g., bone marrow or lymph node). Multiple studies have shown that the presence of these cells is an independent prognostic factor of poor outcome for all tested tumor types [6–8]. For this reason, the genomic characterization of these cells becomes more and more a clinical need. CNAs in single cells have been previously analyzed by Klein et al. [9] by gene metaphase-based comparative genomic hybridization (mCGH) on amplified single-cell genomes. Although the application of this method has

Miodrag Gužvić (ed.), *Single Cell Analysis: Methods and Protocols*, Methods in Molecular Biology, vol. 2752,
https://doi.org/10.1007/978-1-0716-3621-3_11,

proved to be reliable and robust, it had several limitations. Among these, the most important are the low resolution of only 5–10 Mb, as well as the laborious and time-consuming procedure. In this context, aCGH offers a higher sensitivity and specificity [10, 11]. Using arrays composed of highly purified BAC clones, resolutions of 8.3 Mb for single copy deletion could be achieved [12], as well as the detection of aberrant regions as small as 1–2 Mb in cell lines and 4.8 Mb in DCCs [13]. Other studies indicated that using high-density oligonucleotide microarrays the detection limit of single-cell aCGH can be reduced to 1 Mb or less in freshly isolated cells [14, 15]. Here we describe a protocol that enables the detection of aberration down to 100 kb applicable to freshly picked cells, DCCs and CTCs isolated, respectively, from secondary distant site (e.g., bone marrow or lymph nodes) and blood of clinical patients [16]. Additionally, cells derived from other body fluids (e.g., cerebrospinal fluid) or from different isolation procedures, e.g., CellSearch® cartridge, Parsortix™, DEParray™, are well suited for the analysis.

2 Materials

2.1 Reagents

In this protocol high-quality reagents are necessary to execute the aCGH procedure. Utilization of the materials included in the kits, reagents, or instrumentations indicated maximizes the likelihood of a successful experiment. If using third-party materials, the user is recommended to prepare all solutions using ultrapure water (HPLC purified, DNAse-free) and PCR analytical grade reagents.

Store all reagents and enzymes according to the guidelines specified by the manufacturer. Reagent expired or subjected to improper storage may compromise the outcome of the aCGH experiment. Always store the reagents separately from the DNA samples and avoid repeated freeze-thaw cycles of solutions containing DNA or enzymes. When preparing frozen reagent solutions thaw completely, mix briefly on a vortex mixer and spin in a microcentrifuge to collect the contents off the wall and lid, and pre-chill to 4 °C or on ice until use.

2.2 Laboratory Setup

To assure optimal condition for the aCGH, it is recommended to perform all steps under a dedicated vertical laminar flow hood. Make sure that all the pipetting aids are calibrated and handled properly. Always wear powder-free laboratory gloves and use dedicated solutions and pipettes with nuclease-free aerosol-resistant tips to avoid contamination of reagents with nucleases and cross-contaminations of samples. Maintain a clean working area by wiping surfaces with Ethanol 80% or bleaching solutions (e.g., DNA-*Zap*™ or DNA AWAY*) and by UV-irradiation of PCR bench, tips, tubes, and tube racks.

2.3 Safety Notes

Cyanine reagents are considered hazardous and may be harmful if swallowed. Avoid contact with eyes, skin, and clothing.

2× HI-RPM Hybridization Buffers are considered hazardous and may be harmful if swallowed. Avoid contact with eyes, skin, and clothing.

Acetonitrile is considered toxic to blood, kidneys, liver, lungs, gastrointestinal tract, central nervous system, skin, eyes, and upper respiratory tract. Avoid contact with eyes, skin, and clothing.

Always refer to the Material Safety Data Sheet (MSDS) of your specific country.

Carefully follow all waste disposal regulations when disposing waste materials.

2.4 DNA Templates

aCGH application uses a "two-color" process to measure DNA copy number alterations in a *Test Sample* relative to a *Reference Sample.*

1. *Test Sample*: The chapter describes the recommendations to label and hybridize amplified DNA from tumor cells, in particular single DCCs and CTCs, but its spectrum of application is wider and may include other applications where the amount of input DNA material is very limited. Therefore, the method is also well suited for the analysis of cell pools (≥2 cells), specimens extracted from FFPE tissue, as well as cells or DNA material collected with the use of various isolation procedures (e.g., micromanipulators mounted on microscopes, the DEParray™ System). A fundamental prerequisite to the use of this protocol is the amplification with *Ampli 1*™ WGA Kit that has been already described in detail in this book series [17]. Moreover, aCGH of single cells demands for samples with high-quality amplified DNA. Always assess the Genomic Integrity Index (GII) of the samples before proceeding to expensive molecular assays [18]. The use of DNA samples that do not fulfill these quality criteria may result in the failure of the labeling reactions or low quality of the aCGH results (high background noise) that may impair the reliable detection of copy number alterations.
2. *Reference Sample*: A pool of peripheral blood lymphocytes isolated from a healthy male/female donor and amplified with *Ampli 1*™ WGA Kit. Sex-matched or sex-mismatched controls can be used in different settings. If the gender of the sample donor is known, selection of reference DNA is based on the gender of the donor of the test sample. Sex-matched designs are more commonly used when the technology is well established and may offer better sensitivity for detecting gains or losses of regions on the X and Y chromosomes.

Sex-mismatched designs are used when array CGH approaches are first implemented because the dose differences between the X and Y chromosomes provide useful information on assay quality.

3. *Controls*: During procedures generating DNA templates for hybridization, each reaction run should include a no-template control (NTC) where the sample is substituted with PCR grade water.

2.5 Fluorescent Labeling of Amplified DNA

1. Expand Long Template PCR System (Roche) containing: 10× Expand Long Template Buffer 1, 2, and 3 and Expand Long Template Enzyme Mix (5 U/μL).
2. Bovine Serum Albumin from bovine serum, 20 mg/mL (Roche).
3. Fluorescent-labeled nucleotides 25 nmol each (GE-Healthcare): Cy3-dCTP and Cy3-dUTP for the *Reference Sample*, Cy5-dCTP and Cy5-dUTP for the *Test Sample*.
4. 9/10 mixes Deoxynucleotide triphosphate-dNTPs 100 mM each (VWR International): 10 μL dATP, 10 μL dGTP, 9 μL dCTP, 9 μL dTTP, and 62 μL of H_2O.
5. Lib1 primer *5′- TAGTGGGATTCCTGCTGTCAGT -3′* (20 μM): 20 μL of Lib1 (100 μM) and 80 μL of H_2O.

2.6 Purification of the PCR/Digestion Product and Quantification

1. Tru1I (MseI), HC (50 U/μL) restriction enzyme (Thermo-Fisher Scientific).
2. Amicon Ultra-0.5 Centrifugal Filter Unit with Ultracel-100 membrane (Merck).
3. Nanodrop instrument with microarray modus (ThermoFisher Scientific).

2.7 Hybridization

1. Oligo aCGH/ChIP-on-chip Hybridization Kit (Agilent) containing: 2× Hi-RPM Hybridization Buffer and 10× Oligo aCGH/ChIP-on-Chip Blocking Agent.
2. 25% (v/v) Tween20 for molecular biology (Sigma): dissolve 250 μL of Tween20 in PCR grade water.
3. 25% (v/v) Igepal CA-630 for molecular biology (Sigma): dissolve 250 μL of Igepal CA-630 in 750 μL of PCR grade water.
4. Cot Human DNA from human placenta DNA (1 μg/μL), enriched for repetitive sequences (Roche).
5. Thermomixer (Eppendorf).
6. Hybridization Gasket Slides Kit – 4 microarrays per slide format (Agilent) containing gasket slides.

7. SurePrint G3 Human CGH Microarray Kit, 4x180K (Agilent) containing glass slides each formatted with four SurePrint G3 180K arrays.
8. Hybridization Chamber Kit – SureHyb enabled, Stainless (Agilent) containing: stainless steel chamber (top and base), thumb clamp and screw, plastic tweezers, user's guide.
9. Microarray Hybridization Oven (Agilent).

2.8 Washing

1. Slide-staining jar and slide-staining dishes (alternative providers).
2. 50 mL pyrex jar (alternative providers).
3. Oligo aCGH/ChIP-on-chip Wash Buffer Kit (Agilent) containing: Oligo aCGH/ChIP-on-Chip Wash Buffer 1 and Oligo CGH ChIP-on-Chip Wash Buffer 2.
4. Oscillator Horizontal Rotor (alternative providers).
5. Acetonitrile >99.9% (VWR).

2.9 Scanning

1. Ozone Barrier Slide Cover Kit (Agilent).
2. Slide Holders for Agilent DNA Microarray Scanner CA and BA (Agilent).
3. DNA Microarray Scanner (Agilent).
4. Feature Extraction Software (Agilent).
5. Genomic Workbench Software (Agilent).

3 Methods

The overview of this protocol is given in Fig. 1.

3.1 PCR-Based Labeling Using Incorporation of Dye-Conjugated dNTPs and Tru1I Digestion

Purpose In this step *Test* and *Reference samples* are differentially labeled with dUTPs and dCTPs nucleotides conjugated to different Cyanine dyes, Cy5 and Cy3 that have a red and green fluorescence, respectively. Cycle after cycle, the polymerase will incorporate the fluorescent nucleotides into the newly synthetized DNA strand generating amplicons marked with two different colors. The resulting PCR products are then subjected to digestion with Tru1I restriction endonuclease to cleave off the PCR-adaptors prior to the aCGH hybridization, thereby avoiding unspecific cross-hybridization of *Test sample* with *Reference sample* and oligonucleotide probes on the array.

Duration

1. Hands-on time: 45 min.
2. PCR program: 1 h 25 min.
3. Digestion program: 3 h.

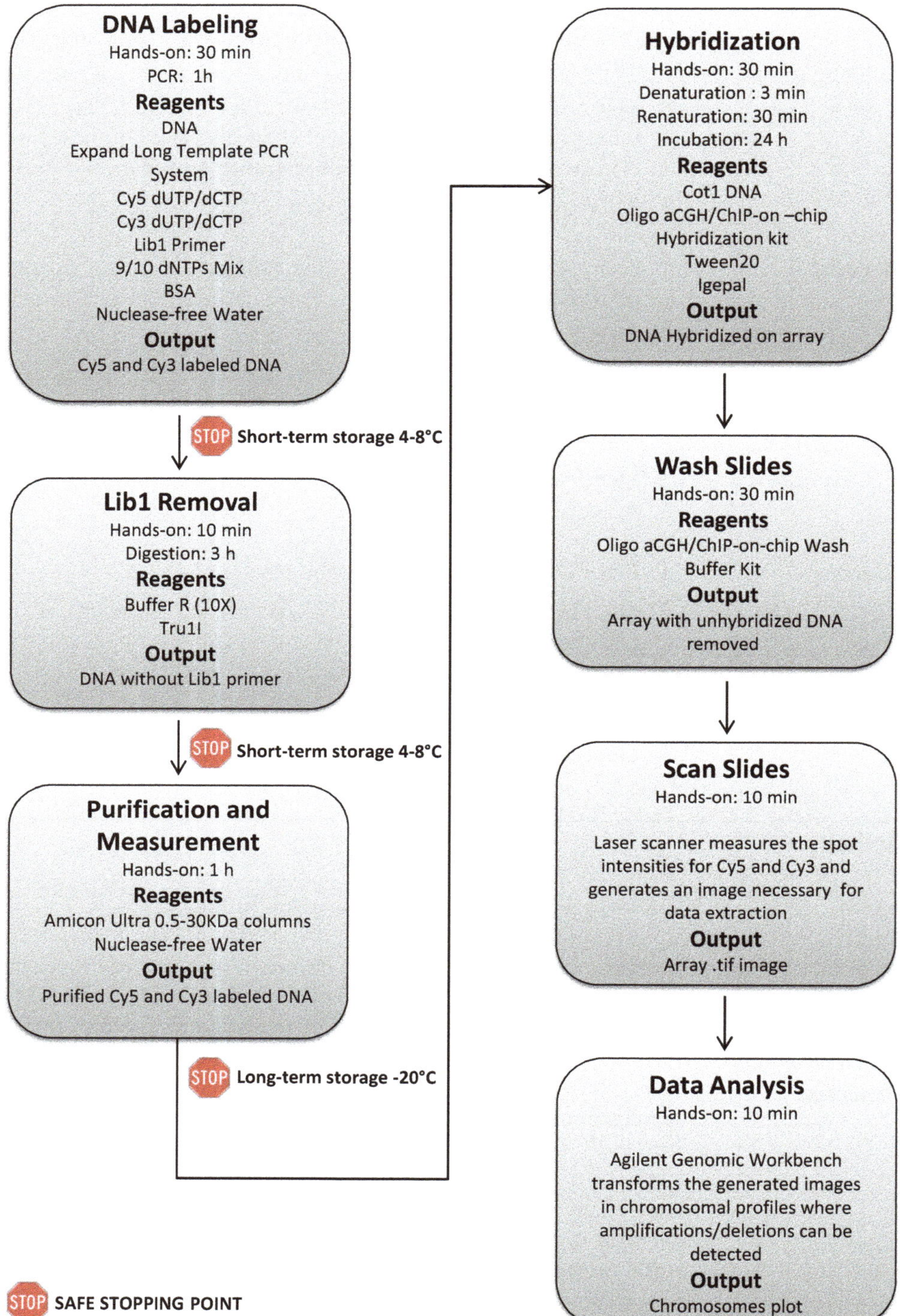

Fig. 1 Overview of the single cell aCGH workflow described in this chapter

Table 1
Cy5/Cy3 labeling master mix

Reagents	Labeling mix Cy3 volume	Labeling mix Cy5 volume
Expand long template buffer 1 (10×)	5.0 μL	5.0 μL
Lib1 (20 μM)	6.0 μL	6.0 μL
9/10 mix dNTPs	1.5 μL	1.5 μL
Cy5 dCTP (25 nmol)	–	1.75 μL
Cy5 dUTP (25 nmol)	–	1.75 μL
Cy3 dCTP (25 nmol)	1.75 μL	–
Cy3 dUTP (25 nmol)	1.75 μL	–
BSA (20 mg/mL)	0.5 μL	0.5 μL
Expand long template DNA polymerase (5 U/μL)	1.5 μL	1.5 μL
DNA template	0.5 μL	0.5 μL
H_2O	31.25 μL	31.25 μL
Total	50.0 μL	50.0 μL

Table 2
PCR program for DNA labeling

Temperature	Time	Cycles
94 °C	15 s	
60 °C	30 s	10 (steps 1–3)
65 °C	3:30 min	
94 °C	15 s	
60 °C	30 s	2 (steps 4–6)
65 °C	3:30 min + 10 s/cycle	
65 °C	7 min	
4 °C	∞	

Procedure (*see* Notes 1 and 2)

1. Prepare the PCR labeling reaction mix (*see* **Notes 3** and **4**): For each *Test* and *Reference sample* set up in parallel two 50 μL PCR reactions according to the recipe outlined in Table 1.
2. Set the PCR program and run the DNA amplification according to the program indicated in Table 2.

Table 3
DNA digestion mix

Reagent	Volume
Buffer R (10×)	5.6 μL
TruII	1.0 μL

Table 4
Thermal profile of DNA digestion

Temperature	Time
65 °C	3 h
4 °C	∞

3. For short-term storage, labeled DNA can be kept at 4–8 °C overnight.
4. Prepare the digestion reaction according to Table 3. Collect the samples from the cycler, briefly centrifuge, and add 6.5 μL of the digestion mix into each labeling reaction. Incubate the DNA digestion using the program outlined in Table 4.
5. For short-term storage, digested DNA can be kept at 4–8 °C overnight.

3.2 Purification of the PCR/Digestion Product and Measurement of the DNA Yield and Incorporation Rate

Purpose In this step the excess labeled nucleotides that have not been incorporated, as well as PCR and digestion reagents, are removed by column purification. Additionally, the cleaved PCR-adaptors generated after digestion are also discarded in the purification process. Only when the samples are purified the measurement can take place on the Nanodrop instrument.

Duration

1. Hands-on time: 30 min.
2. Purification: 30 min.
3. Quantification: 10 min.

Procedure

1. Assemble the Amicon Ultra 0.5 column as described by the manufacturer.
2. Pool the resulting products of the same sample and fill up the volume to 480 μL with H_2O. Centrifuge at 14,000 RCF for 10 min at room temperature.

3. Discard the flow-through and reassemble the purification column assembly. Add 480 μL H_2O and centrifuge again at 14,000 RCF for 10 min at room temperature.
4. Transfer the column in an upside-down (inverted) position to a new 1.5 mL collection tube and centrifuge at 1000 RCF for 1 min at room temperature to elute the purified product (*see* **Note 5**).
5. Discard the purification column and assess the volume of the eluted sample with a pipette. If necessary, fill up the volume to 21 μL with H_2O (*see* **Note 6**).
6. Assess the DNA yield and dye incorporation rates using a NanoDrop instrument.
7. Start the Nanodrop Software and select the "Microarray" modus from the menu.
8. Allow the instrument to perform a self-test, select "DNA-50" from the "Sample Type" menu and blank with H_2O.
9. Select Cy5 and Cy3 dyes for the measurement and keep default absorption/emission settings unchanged.
10. Pipette 2.0 μL of DNA solution directly on the sensor avoiding formation of air bubbles and start the measurement.
11. Wipe the sensor carefully after each measurement with a soft KimTech wipe.
12. Once all the samples have been measured, record the .xls report file in the designated experiment folder.
13. Calculate the

    $\text{Total DNA yield : Concentration DNA} \left(\frac{\text{ng}}{\mu\text{L}}\right) * \text{Sample volume } (\mu\text{L}).$

    Expected yield of the amplified DNA after labeling and clean-up is between 9 and 16 μg.
14. Calculate the Degree of labeling $= \frac{340 * N \left[\frac{\text{pmol}}{\mu\text{L}}\right]\text{dye}}{\text{CDNA} \left[\frac{\text{ng}}{\mu\text{L}}\right] * 1000} * 100\%$

    where N is the dye concentration and CDNA is the DNA concentration.
15. Calculate the Specific activity $= \frac{\text{Degree of labeling}}{0.034}$. The specific activity of Cy3 labeled samples is expected to be more than 8 pmol/μg while for Cy5 more than 11 pmol/μg.
16. Labeled DNA can be stored up to 1 month at −20 °C in the dark.

3.3 Hybridization

Purpose During this step the labeled genomic DNA from the *Test* and the *Reference samples* are mixed in equal amounts and co-hybridized to an array containing immobilized 60-mers of nucleotides. Cot-1 DNA is used to reduce the cross-hybridization of repetitive DNA sequences to the array preventing the nonspecific binding of a labeled probe of interest to the repetitive DNA sequence (*see* **Note 7**).

Table 5
Hybridization mix

Reagent	Volume
Cot1-DNA (1 μg/μL)	5.0 μL
Agilent blocking reagent (10×)	12.0 μL
HI-RPM hybridization buffer (2×)	60.0 μL
Tween20–25% (v/v)	5.2 μL
Igepal – 25%	5.2 μL
Total	87.4 μL

Duration

1. Hands-on time: 15–20 min (*see* **Note 8**).
2. Processing time of the Thermomixer 1: 3 min.
3. Processing time of the Thermomixer 2: 30 min.
4. Processing time of the Agilent Microarray Oven: 24 h.

Procedure

- Step 1: Denaturation /Renaturation of the hybridization solution.

1. Turn on Thermomixer 1 and 2 and set the following parameters: 95 °C for 3 min at 350 rpm and 37 °C for 30 min at 350 rpm, respectively.
2. Turn on the Agilent Microarray Oven and set the temperature to 65 °C (*see* **Note 9**).
3. Thaw the necessary reagents and the desired amount of *Test* and *Reference samples* at room temperature (*see* **Note 10**).
4. Assemble the Hybridization Mix as indicated in Table 5 (*see* **Note 11**).
5. In a new labeled 1.5 mL Eppendorf tube, mix *Test* and *Reference samples* (19 μL each) (*see* **Note 12**).
6. Add 87.4 μL of the Hybridization Mix into the premixed samples for a total of 125.4 μL.
7. Mix gently by pipetting slowly up and down (*see* **Note 11**) and quickly spin to drive contents to the bottom of the reaction tube.
8. Denature the samples in Thermomixer 1: transfer the samples and incubate at 95 °C for 3 min at 350 rpm (*see* **Note 13**).
9. Renature the samples in Thermomixer 2: transfer the samples swiftly from Thermomixer 1 into Thermomixer 2 and incubate at 37 °C for 30 min at 350 rpm (*see* **Note 14**).

10. Remove the samples from the Thermomixer 2 and spin 1 min at 6000 RCF in a minicentrifuge.
11. Samples are ready and must be hybridized immediately. Keep the samples at 37 °C until the Hybridization starts (*see* **Note 15**).

- Step 2: Hybridization on the array slide.

1. Place the base of the Agilent SureHyb Chamber on a clean flat surface.
2. Remove the Gasket Slide from its casing and place it into the Agilent SureHyb Chamber with the rubber seals of the gasket wells facing up. Do not touch the surface of the gasket chambers at any time (*see* **Note 16**).
3. In a drag-and-drop manner dispense 110 μL of the hybridization sample mixture into each gasket well, recording the order of the samples (*see* **Note 17**).
4. Carefully place the Agilent microarray slide on the top of the Gasket Slide with the "Agilent" labeling bar code (active side) facing down and the numeric barcode side facing up. Do not touch the active side of the slide at any time (*see* **Note 18**).
5. Put the SureHyb Chamber Cover onto the gathered slides and slide the clamp assembly on both.
6. Tighten the clamp firmly onto the chamber.
7. Control the assembled chambers for any signs of spillage and stationary air bubbles (*see* **Note 19**).
8. Place the assembled chambers in the rotator rack of the Agilent Microarray Oven, close and set the rotation speed to 20 rpm (*see* **Note 20**).
9. Hybridize at 65 °C for 22–24 h (*see* **Note 21**).
10. For the washing section clean all the necessary glassware equipment before use.
11. Rinse glass slide-staining jar and slide-staining dishes with copious amount of distilled water.
12. Empty out the water collected in the dishes at least five times.
13. Repeat steps until all traces of contaminating material are removed (*see* **Note 22**).
14. It is essential to equilibrate the temperature of the Agilent Oligo aCGH/ChIP-on-Chip Wash Buffer 2 at 37 °C for optimal performance (*see* **Note 23**).
15. Add Agilent Oligo aCGH/ChIP-on-Chip Wash Buffer 2 to a 50 mL pyrex jar and warm overnight in a water bath (or incubator) set to 37 °C.
16. Put a glass slide-staining jar secured with parafilm and warm overnight in a water bath (or incubator) set to 37 °C.

3.4 Washing

Purpose Washing steps are meant to remove the excess of unbound probes or probes loosely attached to non-complementary DNA while leaving the matched probe target intact. Additionally, they clean the corner spots of the main grid – necessary to position correctly the grid template – from contaminating materials (i.e., dust or dirty particles).

Duration

1. Hands-on time: 5 min.
2. Disassembly in Agilent Oligo aCGH/ChIP-on-Chip Wash Buffer 1: 1 min.
3. Agilent Oligo aCGH/ChIP-on-Chip Wash Buffer 1, wash **step 1**: 2 min 30 s.
4. Agilent Oligo aCGH/ChIP-on-Chip Wash Buffer 1, wash **step 2**: 2 min 30 s.
5. Agilent Oligo aCGH/ChIP-on-Chip Wash Buffer 2, wash **step 1**: 30 s.
6. Agilent Oligo aCGH/ChIP-on-Chip Wash Buffer 2, wash **step 2**: 30 s.
7. Acetonitrile wash: 5 s.

Procedure

- Step 1: Disassembly of the Agilent Hybridization Assembly.

1. Fill a slide-staining jar with 50 mL of room temperature Agilent Oligo aCGH/ChIP-on-Chip Wash Buffer 1 used only for the disassembly (*see* **Note 24**).
2. Remove the Agilent Hybridization Assembly from the Agilent Hybridization Oven (*see* **Note 25**).
3. Untighten the clamp and disassemble the Agilent Hybridization Assembly (*see* **Note 26**).
4. Remove the slide sandwich (aCGH and Gasket Slide) from the base of the Agilent SureHyb Chamber and transfer it into the slide-staining jar containing the Agilent Oligo aCGH/ChIP-on-Chip Wash Buffer 1 (*see* **Note 27**).
5. Using a plastic forceps, carefully, peel open the slide sandwich by exerting a slight pressure between the two-stacked glasses (*see* **Note 28**).

- Step 2: Washing Steps (*see* **Note 29**).

1. Before starting the washing steps, under a fume hood, fill a slide-staining dish with Acetonitrile in sufficient amount to cover an aCGH slide.

2. Transfer the aCGH slide into a new slide-staining jar filled with Agilent Oligo aCGH/ChIP-on-Chip Wash Buffer 1 and wash on the horizontal oscillator at room temperature for 2 min and 30 s, 120 rpm (*see* **Note 30**).
3. Invert the aCGH slide and place it back again in the same slide-staining jar, wash on the horizontal oscillator at room temperature for 2 min and 30 s, 120 rpm (*see* **Note 31**).
4. Shortly before the end of the wash prepare the slide-staining jar and the Agilent Oligo aCGH/ChIP-on-Chip Wash Buffer 2 that have been equilibrated overnight at 37 °C.
5. Transfer the aCGH slide into the slide-staining jar filled with Agilent Oligo aCGH/ChIP-on-Chip Wash Buffer 2 and wash on the horizontal oscillator at 37 °C for 30 s, 120 rpm (*see* **Note 27**).
6. Invert the aCGH slide and place it back again in the same slide-staining jar, wash on the horizontal oscillator at 37 °C for 30 s, 120 rpm.
7. Dip the aCGH slide shortly (3–5 s) into the slide-staining dish filled with Acetonitrile.
8. Using a KimTech tissue, wipe gently the remaining drops on the side of the slide (*see* **Note 32**).
9. Place the aCGH slide in the Slide Holder with the "Agilent" side of the slide facing up.
10. Place the Ozone barrier slide on top of the aCGH slide (*see* **Note 33**).
11. Close the Slide Holder and push on the tab end until you hear it click.
12. The aCGH slide is now ready for scanning (*see* **Note 34**).

3.5 Scanning and Extraction

Purpose The aCGH slides are scanned into image files using a specific microarray scanner. During this operation the spot intensities for the two colors are measured and a .tif image, necessary for the data extraction, is generated. In the Feature extraction software, the fluorescence data are translated into log ratio, allowing the identification of aberrations in the *Test Sample*.

Duration

1. Hands-on time: 5 min.
2. Scanning time: Depending on the number of aCGH slides analyzed.
3. Extraction time: Depending on the number of aCGH slides analyzed.

Procedure

- Step 1: Scanning.

1. Switch on the Agilent Microarray Scanner 30 min before to allow the laser to warm up until the light of the scanner turns from “orange” to “green.”
2. Start the program and wait until the instrument is connected to the software.
3. Put the assembled Slide Holder with the Ozone Barrier Slide cover into the scanner carousel always starting from position *number 1*.
4. Select the Start Slot *m* and End Slot *n* representing, respectively, the first and last aCGH slide to be scanned.
5. Select *Profile Agilent G3_aCGH* and verify the following settings:
   - *Dye channel*: R + G (red and green).
   - *Scan region*: Agilent HD (61 × 21.6 mm).
   - *Scan resolution*: 3 μm.
   - *Tiff file dynamic range:* 16 bit.
   - *Red PMT gain:* 100%.
   - *Green PMT gain:* 100%.
   - *XDR:* no XDR.
6. Select the desired location to save the .tif file in the *Output Path Browse*.
7. Verify that the Scanner status in the main window says *Scanner Ready*.
8. Click Scan Slot m-n on the Scan Control main window.

- Step 2: Analyze microarray image.

1. After the scanning is complete, extract the feature with Agilent Feature Extraction Software.
2. A QC metrics file generated after data extraction is used to evaluate the relative data quality of the hybridization set and if potential errors have happened during the analysis (i.e., alignment grid not properly placed).
3. The thresholds that can be observed when analyzing single cells according to this protocol are the following:
   - BGNoise: <10
   - Signal Intensity: >50
   - Signal to Noise: >30
   - Reproducibility: <0.2
   - DLRS: <1

4 Notes

1. Alternative protocol: Random-primed DNA labeling approach. *Test* and *Reference* DNA samples are labeled using SureTag DNA Labeling Kit (Agilent Technologies) according to the instruction provided by the supplier (Agilent Oligonucleotide Array-Based CGH for Single Cell Analysis, August 2015). Briefly, 1.0 μg of the purified input DNA (*see* **Note 3**) diluted in 26 μL of H_2O is supplemented with 5 μL of Random Primer Mix to a final volume of 31 μL. Samples are denatured at 95 °C for 3 min and then kept at 4 °C or on ice. To the 31 μL of denatured DNA add the following ingredients: 10 μL of 5× Reaction Buffer, 5 μL of 10× dNTP Mix, 3 μL of Cy5-dUTP (*Test*) or Cy3-dUTP (*Reference*) and 1 μL of $Exo^{(-)}$ Klenow fragment. Incubate the labeling reaction at 37 °C for 2 h, followed by an inactivation step at 65 °C for 10 min. Purify the labeled DNA using Amicon Ultra 0.5 purification system with a size cut-off of 30 kDa. Quantify DNA yields and dye incorporation rates as described in the main protocol. Typical specific activities are 20–50 pmol/μg for Cyanine 3 and 15–40 pmol/μg for Cyanine 5, with a final DNA yield of 9–15 μg.
2. Take care to keep all the reagents on ice during the reaction preparation, unless otherwise indicated. Return enzymes, dNTPs, and Cyanine nucleotides at −20 °C as soon as possible. Check that the correct program is set in the thermocycler and that evaporation of the samples is avoided by using a heated lid. Make sure that all the pipette aids are calibrated. All these conditions, if improperly managed, can result in a suboptimal specific activity and degree of labeling of the samples.
3. Sometimes the amount of DNA that can be available is limited, especially in the case of clinical samples (CTCs or DCCs). For this reason, the original WGA can be re-amplified for a limited number of cycles in order to generate sufficient amount of DNA (especially needed for the enzymatic labeling, *see* **Note 1**). The short re-amplification does not introduce any significant PCR bias that could affect the reliability of the aCGH in terms of aberrations detection. Run three reactions in parallel for each sample as follows: 5 μL Expand Long Template Buffer 1 (Roche Diagnostic), 1 μM of the Lib1 primer, 1.75 μL dNTPs (10 mM), 1.25 μL BSA (Roche Diagnostic), 2.5 U of Expand-Long-Template DNA Polymerase (Roche Diagnostic), and 1 μL of the template DNA. Set the thermocycler as follows: 1 cycle of 94 °C for 60 s, 60 °C for 30 s, 65 °C for 2 min, 10 cycles of 94 °C for 30 s, 60 °C for 30 s, 65 °C for 2 min (extended by 20 s/cycle). Pool the reactions after

completion and use directly for *PCR-based labeling using incorporation of dye-conjugated dNTPs* as described in the main procedure or, alternatively, purify the reamplified DNA using Amicon Ultra 0.5 purification system with a size cut-off of 30 kDa and proceed with the *Random-primed DNA labeling approach* (*see* **Note 1**).

4. In order to reduce pipetting errors of small volumes of DNA always add 1 μL of DNA to 99 μL of PCR reaction, mix well by short vortexing, and spin down to drive contents off the walls and lid. Divide the reaction into two tubes of 50 μL each. This procedure ensures a proper homogeneous mixing of the DNA-reaction solution minimizing PCR biases in the individual reactions.
5. Remember to change the centrifugation speed to 1000 RCF. If the speed is too high, this will cause perforation of the column membrane and consequent loss of the sample.
6. The volume per sample will be approximately 12–20 μL.
7. The efficient suppression of repeated sequences by the Cot-1 DNA during hybridization represents a relevant factor for the success of array-CGH. Repetitive freeze and thawing of Cot-1 DNA tend to reduce the quality and may alter the suppression efficiency. The use of lower-quality Cot-1 DNA may result in incomplete repeat suppression, altered log ratios, and increased background.
8. All the indicated hands-on processing times were estimated for experiments involving four arrays (e.g., 16 samples in total). Accordingly, hands-on processing times may deviate, when processing higher or lower number of samples in one run. We recommend to get acquainted with the method before processing a higher number of samples. To practice hybridization, prepare a 1:1 2× HI-RPM Hybridization Buffer and water mix and use a microscope slide or used microarray slide, and a gasket slide. You can use the same slide to practice wash and placement of slide in the slide holder.
9. Switch on in advance the Agilent Microarray Oven and the Thermomixers. It may take a while before the exact temperatures will be reached. Check always that the correct temperature is reached and make sure that the oven is calibrated. Too high temperatures of the Agilent Microarray Oven may be responsible for post-labeling signal loss due to hybridization conditions that are too stringent. Too low temperatures may decrease the stringency and increase the BGNoise and DLRSD.
10. For this and all future steps we recommend working in dimmed light and covering your samples with aluminum foil to slow down any possible photo-bleaching.

11. The Hybridization Mix is viscous and foams very intensively due to the presence of Tween20 and Igepal. Avoid to vortex intensively, mix by pipetting slowly up and down or by gently inverting the tube. Care must be taken when pipetting to ensure that the correct volumes are used. Using an incorrect volume will change the stringency of the hybridization solution and could lead to poor probe performance. Always prepare the Hybridization Mix with an excess of one sample.
12. After pairing the Cy5 labeled *Test* and Cy3 labeled *Reference*, the sample should look purple in color. If it is too pink or too blue, it will indicate, respectively, inefficient Cy5 or Cy3 labeling. Inefficient labeling can result from WGA with low GII index, many cycles of freeze-thaw of dNTPs mix and Cyanine nucleotides, light and/or air exposure of the dyes, incorrect volume pipetting, wrong temperatures or times, and degraded enzymes not properly stored on ice.
13. Always protect the samples from direct sunlight exposure by covering the Thermomixer with an aluminum foil. Keep strictly the incubation time because this is a time-sensitive step.
14. The duration of this step can be extended, if necessary, until the moment of the hybridization. Nevertheless, extend only for the necessary time because at 37 °C part of the DNA may start to renature, rendering it unavailable for the hybridization on the array. This process could cause loss of information.
15. Keep the temperature of hybridization sample mixtures as close as possible to 37 °C. To do this, process the samples one by one by keeping them in the Thermomixer 2.
16. Make sure that the Gasket Slide is lying flat at the bottom of the Agilent SureHyb Chamber. Help you in the process by using a plastic forceps to align the slide at one side of the rectangular section of the chamber base. Ensure that the gasket slide is flush with the chamber base and is not unlatched from the borders. This will help to keep the slide stable, avoiding accidental moving, during the dispensing phase of the samples. Keep in mind that this phase is also very sensitive to cross-contamination, especially when samples are loaded onto each of the four adjacent Gasket wells.
17. Start to dispense the samples from the "Subarray 1" (the nearest to the "Agilent" label) to the "Subarray 4" (the most far). Slowly dispense your sample at the center of the well without touching the glass slide. To prevent leakage of the hybridization solution, avoid touching the O-ring of the gasket slide with the hybridization mixture. Proceed with this step swiftly avoiding that the samples loaded on the array are cooling down for too long. Take care to not form too many

bubbles on the array. In addition, avoid pipetting the last drop of Hybridization Mix on the Gasket Slide because it will form a bubble.

18. Hold the Agilent aCGH slide on the sides or on the barcode side only. Approach it, as near as possible, to the Gasket Slide and release when you reach a short distance between them, in order to avoid spreading the Hybridization-Sample mixture outside of the Gasket well and causing material loss and sample cross-contamination. Once the aCGH slide is placed on the Gasket Slide, do not try to readjust the position as this may cause spillage and cross-contamination of samples.
19. Control, by rotating the chamber, if there are stationary bubbles. Knock the Hybridization Assembly with your hand or on a hard surface until the stationary bubbles become mobile or they merge to form a large mobile bubble. It is important that multiple small bubbles are not stationary at the same place. This would compromise the hybridization of the DNA in the affected area of the array.
20. Be always sure that each position in the Agilent Oven is balanced, as you would do for a centrifuge. Use an empty assembled Hybridization Chamber to balance the loaded Hybridization Chambers on the rack to prevent unnecessary overburden on the oven motor.
21. The hybridization step can be shortened or elongated according to the necessity. Do not hybridize less than 20 h because this may impact the quality of the hybridization.
22. Make sure that the staining jar and dishes are very clean. High BGNoise values can be caused by glassware not properly washed. Always use MilliQ water to wash equipment, do not use tap water or any detergents. If high BGNoise persists, the glassware should be washed with acetonitrile followed by rinses with MilliQ water to remove build-up of unincorporated dye molecules.
23. Post-labeling signal loss of Cy5 can be due to the incorrect temperature of Agilent Oligo aCGH/ChIP-on-Chip Wash Buffer 2. Always ensure that the temperature is set correctly at 37 °C.
24. We recommend preparing a staining jar filled with wash buffer for each sample or, alternatively, washing the staining jar in between each array disassembly with copious amounts of MilliQ water and refill with fresh buffer. In this way, cross-contamination between slides can be minimized.
25. We recommend washing array slides one by one. If there are arrays left in the Hybridization Oven, keep them until you are finished washing the first slide. Control the level of the fluid among the four Gasket Chambers to ensure that there was no

loss of the samples due to leakage. Cross-contamination can arise during the overnight incubation due to gasket leakage. If you notice a loss of fluid, you may consider that the presence of contamination can result in altered and inaccurate patient data.

26. This step may take 2–3 min because the Hybridization Assembly is too hot and it is not possible to untighten the clamp. Wait until it reaches a lower temperature to avoid burns of the hands.
27. We recommend wearing multiple layers of gloves that can be quickly changed between each step. Change them after the disassembly of the Hybridization Assembly as you can notice a black smear on your gloves. Change your gloves when you step from Agilent Oligo aCGH/ChIP-on-Chip Wash Buffer 1 to Wash Buffer 2 and to Acetonitrile. This procedure will avoid transporting traces of contaminants during the different washing steps.
28. Do not rush at this stage. Exert a light pressure to avoid the harsh separation of the Gasket Slide from the aCGH slide that may cause the loss of the DNA-probe complex from the surface of the array and/or the break of the glass. Let the Agilent Oligo aCGH/ChIP-on-Chip Wash Buffer 1 slowly penetrate between the two slides. This will reduce adherence and facilitate the separation.
29. Wash slides in an ozone-controlled environment. Ozone can affect the fluorescent signal during the washing step of the arrays. You may notice Cy5 signal red signal loss around the edges of the feature because after the last wash buffer, the outer edges will dry before the center, and, as a consequence, they will become exposed earlier to the degrading effect of ozone. Signal degradation can potentially compromise data quality. We recommend using small batches of slides that can be washed and scanned in a short time to minimize exposure to air.
30. Keep strictly the incubation time and rpm of the Horizontal Oscillator. High BGNoise is often introduced if uncorrected wash steps are performed. The indicated rpm is necessary to remove all non-uniformities from the surface of the array and excess of unbound labeled DNA.
31. By inverting the aCGH slide, it is ensured a more homogeneous wash throughout the entire array, in the upper and in the lower part.
32. Do not wipe with the KimWipe the active surface of the aCGH slide (Agilent site) if there are residual drops. Rather, submerge again shortly after the aCGH slide in Acetonitrile. Limit yourself to clean only the edges from eventual small drops. In general, Acetonitrile is very volatile and it does not leave the aCGH slide wet, indeed it leaves very dry.

33. The second stage where ozone can cause a problem occurs when the slide is being scanned. The increased air circulation at the open end of the slide holder (left side) causes more ozone to contact the Cy5 thereby inducing a visible degradation gradient. To prevent degradation during scanning we recommend using a barrier slide, which sits over the microarray. This seals the array in a small chamber and limits the extent of ozone degradation.
34. Scanning the slides immediately after this step is highly recommended to minimize the reduction in the fluorescence signal intensity due to environmental oxidants and possible photobleaching. In case immediate scanning is not possible keep the slides in black light-sealed boxes (you can use the boxes the array slides come in) or slide-holding boxes wrapped in aluminum foil.

References

1. Pinkel D, Segraves R, Sudar D, Clark S, Poole I, Kowbel D, Collins C, Kuo WL, Chen C, Zhai Y, Dairkee SH, Ljung BM, Gray JW, Albertson DG (1998) High resolution analysis of DNA copy number variation using comparative genomic hybridization to microarrays. Nat Genet 20(2):207–211. https://doi.org/10.1038/2524
2. Solinas-Toldo S, Lampel S, Stilgenbauer S, Nickolenko J, Benner A, Dohner H, Cremer T, Lichter P (1997) Matrix-based comparative genomic hybridization: biochips to screen for genomic imbalances. Genes Chromosomes Cancer 20(4):399–407
3. Fiorentino F (2012) Array comparative genomic hybridization: its role in preimplantation genetic diagnosis. Curr Opin Obstet Gynecol 24(4):203–209. https://doi.org/10.1097/GCO.0b013e328355854d
4. Munne S (2012) Preimplantation genetic diagnosis for aneuploidy and translocations using array comparative genomic hybridization. Curr Genomics 13(6):463–470. https://doi.org/10.2174/138920212802510457
5. Fiorentino F, Caiazzo F, Napolitano S, Spizzichino L, Bono S, Sessa M, Nuccitelli A, Biricik A, Gordon A, Rizzo G, Baldi M (2011) Introducing array comparative genomic hybridization into routine prenatal diagnosis practice: a prospective study on over 1000 consecutive clinical cases. Prenat Diagn 31(13):1270–1282. https://doi.org/10.1002/pd.2884
6. Klein CA (2003) The systemic progression of human cancer: a focus on the individual disseminated cancer cell--the unit of selection. Adv Cancer Res 89:35–67
7. Riethdorf S, Wikman H, Pantel K (2008) Review: biological relevance of disseminated tumor cells in cancer patients. Int J Cancer 123(9):1991–2006. https://doi.org/10.1002/ijc.23825
8. Cohen SJ, Punt CJ, Iannotti N, Saidman BH, Sabbath KD, Gabrail NY, Picus J, Morse M, Mitchell E, Miller MC, Doyle GV, Tissing H, Terstappen LW, Meropol NJ (2008) Relationship of circulating tumor cells to tumor response, progression-free survival, and overall survival in patients with metastatic colorectal cancer. J Clin Oncol 26(19):3213–3221. https://doi.org/10.1200/JCO.2007.15.8923
9. Klein CA, Schmidt-Kittler O, Schardt JA, Pantel K, Speicher MR, Riethmuller G (1999) Comparative genomic hybridization, loss of heterozygosity, and DNA sequence analysis of single cells. Proc Natl Acad Sci U S A 96(8):4494–4499
10. Greshock J, Feng B, Nogueira C, Ivanova E, Perna I, Nathanson K, Protopopov A, Weber BL, Chin L (2007) A comparison of DNA copy number profiling platforms. Cancer Res 67(21):10173–10180. https://doi.org/10.1158/0008-5472.CAN-07-2102
11. Haraksingh RR, Abyzov A, Gerstein M, Urban AE, Snyder M (2011) Genome-wide mapping of copy number variation in humans: comparative analysis of high resolution array platforms. PLoS One 6(11):e27859. https://doi.org/10.1371/journal.pone.0027859

12. Fiegler H, Geigl JB, Langer S, Rigler D, Porter K, Unger K, Carter NP, Speicher MR (2007) High resolution array-CGH analysis of single cells. Nucleic Acids Res 35(3):e15. https://doi.org/10.1093/nar/gkl1030
13. Fuhrmann C, Schmidt-Kittler O, Stoecklein NH, Petat-Dutter K, Vay C, Bockler K, Reinhardt R, Ragg T, Klein CA (2008) High-resolution array comparative genomic hybridization of single micrometastatic tumor cells. Nucleic Acids Res 36(7):e39. https://doi.org/10.1093/nar/gkn101
14. Bi W, Breman A, Shaw CA, Stankiewicz P, Gambin T, Lu X, Cheung SW, Jackson LG, Lupski JR, Van den Veyver IB, Beaudet AL (2012) Detection of >/=1 Mb microdeletions and microduplications in a single cell using custom oligonucleotide arrays. Prenat Diagn 32(1):10–20
15. Mohlendick B, Bartenhagen C, Behrens B, Honisch E, Raba K, Knoefel WT, Stoecklein NH (2013) A robust method to analyze copy number alterations of less than 100 kb in single cells using oligonucleotide array CGH. PLoS One 8(6):e67031. https://doi.org/10.1371/journal.pone.0067031
16. Czyz ZT, Hoffmann M, Schlimok G, Polzer B, Klein CA (2014) Reliable single cell array CGH for clinical samples. PLoS One 9(1):e85907. https://doi.org/10.1371/journal.pone.0085907
17. Czyż ZT, Klein CA (2015) Deterministic whole-genome amplification of single cells. In: Kroneis T (ed) Whole genome amplification methods and protocols. Springer, New York
18. Polzer B, Medoro G, Pasch S, Fontana F, Zorzino L, Pestka A, Andergassen U, Meier-Stiegen F, Czyz ZT, Alberter B, Treitschke S, Schamberger T, Sergio M, Bregola G, Doffini A, Gianni S, Calanca A, Signorini G, Bolognesi C, Hartmann A, Fasching PA, Sandri MT, Rack B, Fehm T, Giorgini G, Manaresi N, Klein CA (2014) Molecular profiling of single circulating tumor cells with diagnostic intention. EMBO Mol Med 6(11):1371–1386. https://doi.org/10.15252/emmm.201404033

Chapter 12

Single Cell Micro RNA Sequencing Library Preparation

Sarah M. Hücker and Stefan Kirsch

Abstract

Micro RNAs represent important post-transcriptional regulators in health and are involved in the onset of many diseases. Therefore, the further characterization of physiological miRNA functions is an important basic research question, and miRNAs even have high potential as biomarkers both for prognosis and diagnosis. In order to exploit this potential, it is mandatory to accurately quantify the miRNA expression not only in bulk but also on the single-cell level. Here, we describe a protocol, which facilitates miRNA sequencing library preparation of very low input samples, single cells, and even clinical samples such as circulating tumor cells. The protocol can be combined with different single-cell isolation methods (e.g., micromanipulation and FACS sorting). After cell lysis, sequencing adapters are ligated to the miRNAs, other ncRNA species, and adapter dimers are reduced by exonuclease digest, the miRNA library is reverse transcribed, amplified, and purified. Furthermore, quality controls are described to select only high-quality samples for sequencing.

Key words miRNA, small RNA, ncRNA, Single-cell sequencing, NGS, CTC

1 Introduction

Micro RNAs govern mRNA abundance and represent important post-transcriptional regulators in many biological processes [1]. The expression levels of miRNAs are very heterogeneous between developmental stages and different tissue or cell types [2]. Also, in diseases like cancer the expression level of some miRNAs is altered [3, 4]. In order to investigate the role of miRNAs in health and disease in further detail, accurate detection and quantification methods are required. Compared to mRNA sequencing, miRNA sequencing poses some additional challenges: First, miRNAs neither have a polyA tail nor any other specific sequence feature, which discriminates them from other RNA types inside the cell. Second, mature miRNAs have a size of only 20–24 nt, which hinders the discrimination and separation of adapter dimers from libraries with insert. Ideally, the library preparation method already avoids the formation of adapter dimers. Third, the quantification accuracy is only moderate due to ligation and amplification

Miodrag Gužvić (ed.), *Single Cell Analysis: Methods and Protocols*, Methods in Molecular Biology, vol. 2752,
https://doi.org/10.1007/978-1-0716-3621-3_12,

biases [5–8]. For bulk miRNA sequencing, several commercial kits and protocols are available [9, 10]. However, for single-cell miRNA sequencing only very few protocols have been published [11–15]. Therefore, we compared 19 miRNA sequencing protocol variants and selected the best-performing protocol regarding the amount of adapter dimers, the amount of reads mapping to annotated miRNA loci, the number of different detected miRNAs, the reproducibility, and the quantification accuracy. The best-performing protocol showed reliable results with very low amounts of miRNA input, single cells of different cell lines, and circulating tumor cells from the blood of small cell lung cancer patients. In addition, the protocol captured other non-coding RNAs like lncRNAs, rRNAs, snoRNAs, and piRNAs [16]. The protocol can be performed in 2–3 days and the hands-on-time is about 8 h. Up to 20 samples can be processed in parallel. First, reaction vessels with lysis buffer are prepared. Then, a single-cell suspension of the target cells has to be made. Single cells are picked using micromanipulation into the lysis buffer and could be stored at −80 °C for up to 6 months. During the library preparation, cells are lysed, 5.8S rRNA is depleted, chemically modified adapters are ligated to the 3′ and 5′ miRNA ends, remaining unligated adapter is removed, reverse transcription is performed, and the cDNA is amplified in two sequential PCRs. In order to enrich for fragments with miRNA inserts, a bead size selection is performed. Finally, the concentration and fragment length distribution of the library is controlled.

Figure 1 summarizes the protocol workflow.

2 Materials

To avoid sample contamination or RNA degradation, only sterile, DNase- and RNase-free reaction tubes should be used, and all pipetting steps should be conducted using sterile filter/barrier tips. Furthermore, all plastic labware should be low retention to reduce sample loss.

2.1 Oligonucleotides

All oligonucleotides should be purified by HPLC. The lyophilized oligonucleotides are dissolved in nuclease-free water to a stock concentration of 100 μM. DNA oligonucleotides can be stored at −20 °C. RNA oligonucleotides are aliquoted and stored at −80 °C. The aliquots are thawed only once and then discarded. The names and sequences of the required oligonucleotides are listed in Table 1.

2.2 Reagents

If not stated differentially, the reagents are stored at −20 °C and thawed on ice.

1. 1% Triton X-100 (Sigma-Aldrich) in nuclease-free water, store at RT.

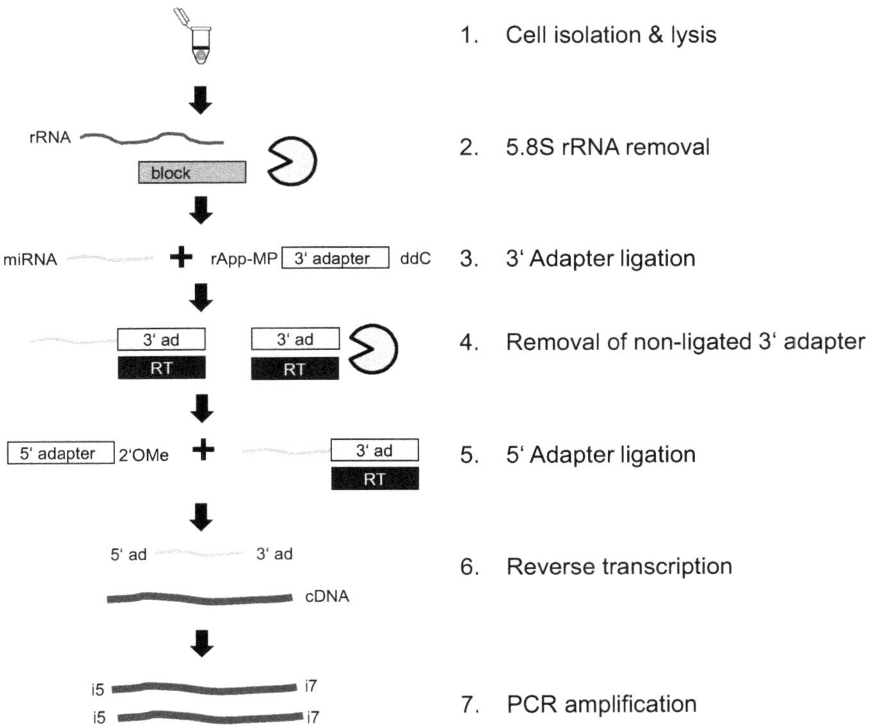

Fig. 1 Workflow of the single-cell miRNA sequencing protocol

Table 1
Names and sequences of the required oligonucleotides

Name	Sequence (5′ → 3′)	Note
3′ adapter	rApp-T(MP)GGAATTCTCGGGTGCCAAGG-ddC	rApp = 5′ preadenylation (*see* **Note 1**) MP = methyl-phosphonate ddC = dideoxy-cytidine
5′ adapter	rGrUrUrCrArGrArGrUrUrCrUrArCrArGrUrCrCrG rArCrGrArUrCrHrHrHrHrHrHrHrHrCrA-2′OMe	r = ribonucleotides H = rA, rC or rU 2′OMe = ribose 2′ O methylation
RT primer	Biotin-CCTTGGCACCCGAGAATTCCrA	r = ribonucleotides
5.8S rRNA mask	ATCGGCAAGCGACGCTCAGACAGGCGTAG CCCCGGGAGGAACCCGGGGCCGCAAGTGC GTTCGAAGTGTCGATGAT-TEG-biotin	TEG = triethyleneglycol spacer
RP1 primer	AATGATACGGCGACCACCGAGATCTACAC GTTCAGAGTTCTACAGTCCGA	
Index-RPI 1–48	CAAGCAGAAGACGGCATACGAGATXXXXXX GTGACTGGAGTTCCTTGGCACCCGAGAATTC CA	XXXXXX = index sequence 1–48, use the indices of the Illumina TruSeq small RNA library preparation kit

2. 40 U/μL Recombinant RNase Inhibitor (Takara).
3. 200 U/μL T4 RNA Ligase 2 truncated KQ, 10× Reaction Buffer and 50% w/v PEG 8000 (NEB).
4. 5 U/μL Lambda Exonuclease (NEB).
5. 50 U/μL 5′ Deadenylase (NEB).
6. 10 U/μL T4 RNA Ligase 1 ssRNA Ligase (NEB).
7. 200 U/μL SuperScript II Reverse Transcriptase and 100 mM DTT (Invitrogen).
8. 2 U/μL Phusion Hot Start II DNA Polymerase and 5× Phusion HF Buffer (Thermo Fisher).
9. 100 mM Tris-buffered ATP (Thermo Fisher).
10. 10 mM each PCR Deoxynucleotide Mix (Roche).
11. 10× Taq DNA Polymerase Buffer (Roche).
12. Ampure XP Beads (Beckman Coulter), store at 4 °C.
13. 80% ethanol solution, store at 4 °C.
14. Nuclease-free water, store at RT.
15. Phosphate buffered saline, PBS, store at RT.
16. Bioanalyzer High Sensitivity DNA Kit (Agilent).
17. Qubit™ 1× dsDNA HS Assay Kit (Invitrogen).

2.3 Equipment

1. Inverted microscope.
2. Micromanipulator Patchman with CellTram pump (Eppendorf).
3. AdcellTM diagnostic slides (Thermo Fisher).
4. Micro Hematocrit capillary (Brand).
5. DMZ Universal Puller (Zeitz Instruments).
6. Thermo cycler.
7. 2100 Bioanalyzer Instrument (Agilent).
8. Qubit 2.0 Fluorometer (Invitrogen).

3 Methods

Single-cell miRNA-Seq is very sensitive to contamination with environmental miRNA or to RNA degradation caused by environmental RNases. Therefore, the experimenter should wear a lab coat and gloves. All pipetting steps should be conducted under a laminar flow hood. Ideally, the hood and the lab equipment (e.g., tube racks, ice boxes) are not commonly used to process amplified nucleic acids.

The protocol Subheadings 3.1, 3.2, and 3.3 should be performed on ice. Subheading 3.4 is performed at room temperature.

The required reagent volumes for one sample are given. If more samples are processed, the volumes have to be increased respectively including some excess. We do not recommend to process more than 20 samples in parallel (*see* **Note 2**).

3.1 Lysis Buffer Preparation

1. Prepare a 1% Triton X-100 solution with water.
2. Prepare the lysis buffer master mix: 0.39 μL 1% Triton X-100, 0.1 μL 40 U/μL RNase inhibitor, and 1.51 μL water.
3. Vortex lysis buffer master mix, pipet 2 μL per sample into 0.2 mL reaction tubes, and spin down.
4. Store tubes with lysis buffer on ice.

3.2 Single Cell Isolation

1. Harvest cells and resuspend cell pellet in PBS (*see* **Notes 3** and **4**).
2. Pick single cells by micromanipulation (*see* **Notes 5** and **6**).
3. Aspirate single cell into 1 μL PBS, pipet into lysis buffer, and centrifuge down.
4. Store the tubes with the isolated cell on dried ice while isolating additional single cells.
5. Long-term storage of isolated cells in lysis buffer is possible at −80 °C.

3.3 miRNA-Seq Library Preparation

1. Process samples from Subheading 3.2 immediately or thaw in ice samples that were stored at −80 °C.
2. Dilute 5.8S rRNA Mask oligo from 100 μM to 5 μM using water.
3. Pipet 2 μL of 5 μM 5.8S rRNA Mask oligo to the sample and spin down (**Note 7**).
4. Incubate at 72 °C for 20 min in a thermo cycler.
5. Prepare the 3′ Adapter ligation mix: 0.16 μL 100 μM 3′ Adapter oligo, 1.28 μL 50% (w/v) PEG 8000, 0.64 μL 10× T4 RNA Ligase buffer, 0.25 μL T4 RNA Ligase 2 truncated KQ, 0.1 μL 40 U/μL RNase inhibitor, and 0.57 μL water.
6. Mix 3′ Adapter ligation mix, pipet 3 μL to the sample and spin down (*see* **Note 8**).
7. Incubate at 30 °C for 6 h, then at 4 °C for 10 h (overnight) in a thermo cycler.
8. Prepare the adapter removal mix: 0.5 μL 100 μM RT primer, 0.5 μL 5 U/μL Lambda Exonuclease, 0.5 μL 50 U/μL 5′ Deadenylase, and 0.5 μL water.

9. Mix adapter removal mix, pipet 2 μL to the sample and spin down.
10. Incubate at 30 °C for 15 min, then at 37 °C for 15 min in a thermo cycler.
11. Prepare the 5′ Adapter ligation mix: 0.12 μL 100 μM 5′ Adapter oligo, 0.08 μL 100 μM ATP, 0.3 μL 10× T4 RNA Ligase buffer, 1.35 μL 10 U/μL T4 RNA Ligase 1 ssRNA Ligase, and 0.15 μL water.
12. Mix 5′ Adapter ligation mix, pipet 2 μL to the sample, and spin down.
13. Incubate at 37 °C for 1 h in a thermo cycler.
14. Prepare reverse transcription mix: 2.2 μL 10× Taq DNA Polymerase Buffer, 1.35 μL 100 mM DTT, 0.85 μL 10 mM dNTPs, 0.5 μL 200 U/μL SuperScript II Reverse Transcriptase, and 0.1 μL 40 U/μL RNase Inhibitor.
15. Mix reverse transcription mix, pipet 5 μL to the sample, and spin down.
16. Incubate at 42 °C for 1 h, then at 70 °C for 15 min in a thermo cycler.
17. Prepare first PCR mix: 0.3 μL 100 μM RP1 primer, 0.45 μL 10 mM dNTPs, 6 μL 5× Phusion HF Buffer, 0.5 μL 2 U/μL Phusion Hot Start II DNA Polymerase, and 5.75 μL water.
18. Mix first PCR mix, pipet 13 μL to the sample, and spin down.
19. Perform the PCR program shown in Table 2 in a thermo cycler.
20. Prepare the indexing PCR mix: 0.2 μL 100 μM RP1 primer, 0.5 μL 10 mM dNTPs, 5 μL 5× Phusion HF Buffer, 0.25 μL 2 U/μL Phusion Hot Start II DNA Polymerase, and 17.55 μL water.

Table 2
Program of the first PCR

Step	Temperature (°C)	Time (s)	Additional
1	98	30	
2	98	10	
3	60	30	
4	72	30	
5			Go to **step 2**, 12 times
6	72	300	
7	4	Forever	

Table 3
Program of the indexing PCR

Step	Temperature (°C)	Time (s)	Additional
1	98	30	
2	98	10	
3	67	30	
4	72	30	
5			Go to **step 2**, 12 times
6	72	300	
7	4	Forever	

21. Mix indexing PCR mix, pipet 23.5 μL into a fresh 0.2 mL reaction tube, and spin down.
22. Mix the sample after first PCR (**step 19**) and pipet 1 μL each into the indexing PCR mix.
23. Pipet 0.5 μL of 100 μM Index RPI oligo into the indexing PCR mix. Use a different Index RPI oligo for every sample (*see* **Note 9**).
24. Mix sample, spin down, and perform the PCR program shown in Table 3 in a thermo cycler.

3.4 Bead Size Selection

1. Bring Ampure XP Beads to room temperature.
2. Vortex beads, pipet 25 μL of beads to 25 μL sample after the indexing PCR, and mix by pipetting up and down 10 times (*see* **Note 10**).
3. Incubate 10 min at RT.
4. Place tube into a magnetic rack and incubate 4 min.
5. Pipet the supernatant into a fresh 0.2 mL reaction tube (*see* **Note 11**).
6. Vortex beads, pipet 15 μL of beads to the supernatant and mix by pipetting up and down 10 times.
7. Incubate 10 min at RT.
8. Place tube into a magnetic rack and incubate 4 min.
9. Remove and discard supernatant.
10. Pipet 200 μL of 80% ethanol to every sample and incubate 30 s (*see* **Note 12**).
11. Remove and discard supernatant.
12. Repeat **steps 10** and **11** once.
13. Remove any residual liquid with a small pipetting tip.

14. Incubate with open lid for 2 min (**Note 13**).
15. Take tube from magnetic rack and resuspend beads in 16 μL of water by pipetting up and down.
16. Incubate 2 min at RT.
17. Place tube into a magnetic rack and incubate 4 min.
18. Pipet 15 μL of the supernatant into a fresh 0.2 mL reaction tube and discard beads.
19. The sample can be stored at −20 °C for at least 1 year.

3.5 Library Quality Control

1. Determine the dsDNA concentration using the Qubit™ 1× dsDNA HS Assay Kit: Prepare standards (10 μL standard + 190 μL Qubit working solution) and samples (1 μL sample + 199 μL Qubit working solution), vortex, incubate for 2 min in the dark, and measure dsDNA concentration (**Note 14**).
2. If the DNA concentration is higher than 2 ng/μL, dilute 1 μL sample to 2 ng/μL with water. If the DNA concentration is lower than 2 ng/μL, just use 1 μL undiluted sample.
3. Determine the fragment length distribution on a Bioanalyzer using a High Sensitivity DNA Kit. Pipet 1 μL of (diluted) sample on the chip (*see* **Notes 15–17**).

4 Notes

1. An already preadenylated 3′ Adapter can be ordered or the preadenylation can be performed using the 5′ DNA Adenylation Kit (NEB).
2. We recommend to include two controls: A positive control consisting of a cell pool or a synthetic miRNA standard (e.g., 10 pg miRXplore Universal Reference from Miltenyi) and a negative control without a cell to detect potential contaminations.
3. The protocol is working not only with cultured cell lines but also with single cells of body fluids and tissues. In this case, the target cell population has to be enriched and converted into a single-cell suspension. It is important to check in advance if these additional processing steps influence RNA quality (e.g., total RNA could be isolated and the RNA integrity score could be determined on the Bioanalyzer).
4. The time between cell harvest and the transfer into lysis buffer should be as short as possible because long processing times could influence miRNA expression.

5. If a fluorescence microscope is used, staining of surface markers could be applied to identify target cells. It should be controlled in advance so that the staining protocol does not influence RNA quality.
6. Other methods than micromanipulation can be used for single-cell isolation (e.g., flow cytometry or single-cell printing). However, the total volume after cell isolation has to be exactly 3 μL.
7. Always pipet on the edge of the tube at the border of the liquid. Do not dip the pipetting tip into the liquid, this can lead to sample loss.
8. PEG 8000 is very viscous. It should be pipetted very slowly and the tip should be controlled for air bubbles and remaining liquid on the outside. The master mix should be mixed at least 30 s and controlled for homogenous appearance of the liquid. Mixing by pipetting up and down might be required to get a homogenous liquid.
9. If the samples should be sequenced together, choose compatible indices regarding the Illumina pooling guide.
10. Do not vortex, only mix beads by pipetting up and down.
11. Control the supernatant in the pipetting tip for beads. If brown beads are visible, pipet the whole liquid back into the tube, incubate again on the magnet, and repeat the removal of the supernatant.
12. The 80% ethanol solution should be prepared freshly every time.
13. The beads should not dry completely, and on the other hand, there should be no remaining liquid visible.
14. The dsDNA concentration varies between different cell types. Libraries of single cells usually have concentrations of 0.5–10 ng/μL.
15. A high-quality library should show the main product at 140–155 bp. Products around 125 bp indicate adapter dimers and fragments larger than 155 bp are caused by nucleic acids other than miRNAs (Fig. 2).
16. We recommend to perform qPCR quantification of the libraries before sequencing. The qPCR is an additional quality control (the Cp value should be lower than 20) as well as necessary for equimolar library pooling.
17. The miRNA libraries can be sequenced on any Illumina sequencer. In the sample sheet, the Illumina TruSeq Small RNA assay and index adapters have to be selected. A read length of 50 nt is sufficient. We usually pool 25–40 libraries for one MiSeq run.

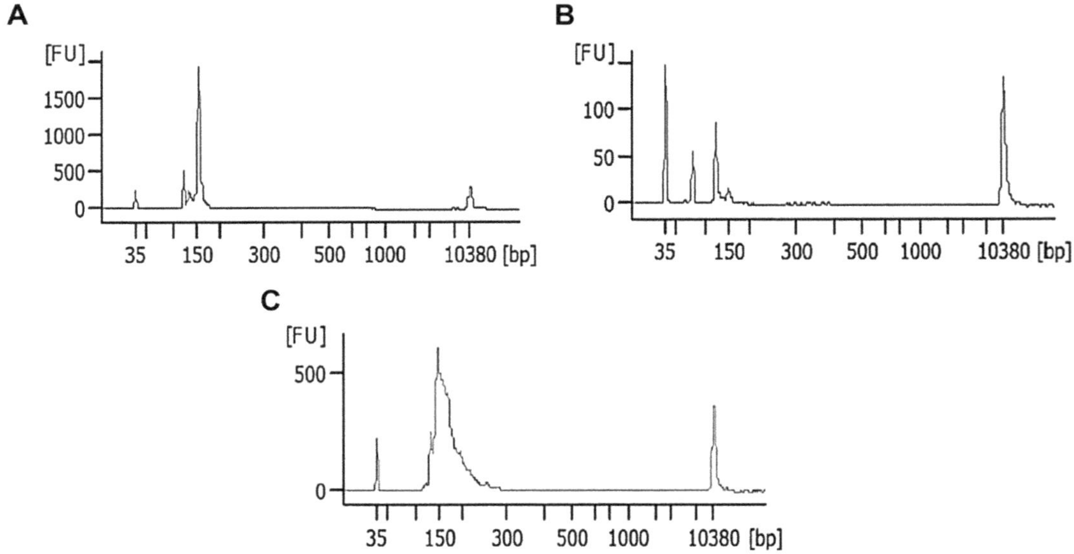

Fig. 2 Example Bioanalyzer profiles of three miRNA libraries. (**a**) This library shows the expected fragment length distribution with the main product at about 150 bp. (**b**) This library contains a large amount of adapter dimers visible by the main product at about 120 bp. (**c**) The main product is larger than 150 bp indicating contamination by larger RNA classes

References

1. Cai Y, Yu X, Hu S, Yu J (2009) A brief review on the mechanisms of miRNA regulation. Genomics Proteomics Bioinformatics 7(4): 147–154. https://doi.org/10.1016/S1672-0229(08)60044-3
2. Ludwig N, Leidinger P, Becker K, Backes C, Fehlmann T, Pallasch C, Rheinheimer S, Meder B, Stahler C, Meese E, Keller A (2016) Distribution of miRNA expression across human tissues. Nucleic Acids Res 44(8): 3865–3877. https://doi.org/10.1093/nar/gkw116
3. Di Leva G, Garofalo M, Croce CM (2014) MicroRNAs in cancer. Annu Rev Pathol 9: 287–314. https://doi.org/10.1146/annurev-pathol-012513-104715
4. Horita M, Farquharson C, Stephen LA (2021) The role of miR-29 family in disease. J Cell Biochem 122(7):696–715. https://doi.org/10.1002/jcb.29896
5. Baran-Gale J, Kurtz CL, Erdos MR, Sison C, Young A, Fannin EE, Chines PS, Sethupathy P (2015) Addressing bias in small RNA library preparation for sequencing: a new protocol recovers microRNAs that evade capture by current methods. Front Genet 6:352. https://doi.org/10.3389/fgene.2015.00352
6. Fuchs RT, Sun Z, Zhuang F, Robb GB (2015) Bias in ligation-based small RNA sequencing library construction is determined by adaptor and RNA structure. PLoS One 10(5): e0126049. https://doi.org/10.1371/journal.pone.0126049
7. Giraldez MD, Spengler RM, Etheridge A, Godoy PM, Barczak AJ, Srinivasan S, De Hoff PL, Tanriverdi K, Courtright A, Lu S, Khoory J, Rubio R, Baxter D, Driedonks TAP, Buermans HPJ, Nolte-'t Hoen ENM, Jiang H, Wang K, Ghiran I, Wang YE, Van Keuren-Jensen K, Freedman JE, Woodruff PG, Laurent LC, Erle DJ, Galas DJ, Tewari M (2018) Comprehensive multi-center assessment of small RNA-seq methods for quantitative miRNA profiling. Nat Biotechnol 36(8): 746–757. https://doi.org/10.1038/nbt.4183
8. Wright C, Rajpurohit A, Burke EE, Williams C, Collado-Torres L, Kimos M, Brandon NJ, Cross AJ, Jaffe AE, Weinberger DR, Shin JH (2019) Comprehensive assessment of multiple biases in small RNA sequencing reveals significant differences in the performance of widely used methods. BMC Genomics 20(1):513. https://doi.org/10.1186/s12864-019-5870-3

9. Dard-Dascot C, Naquin D, d'Aubenton-Carafa Y, Alix K, Thermes C, van Dijk E (2018) Systematic comparison of small RNA library preparation protocols for next-generation sequencing. BMC Genomics 19(1):118. https://doi.org/10.1186/s12864-018-4491-6
10. Herbert ZT, Thimmapuram J, Xie S, Kershner JP, Kolling FW, Ringelberg CS, LeClerc A, Alekseyev YO, Fan J, Podnar JW, Stevenson HS, Sommerville G, Gupta S, Berkeley M, Koeman J, Perera A, Scott AR, Grenier JK, Malik J, Ashton JM, Pivarski KL, Wang X, Kuffel G, Mesa TE, Smith AT, Shen J, Takata Y, Volkert TL, Love JA, Zhang Y, Wang J, Xuei X, Adams M, Levine SS (2020) Multisite evaluation of next-generation methods for small RNA quantification. J Biomol Tech 31(2):47–56. https://doi.org/10.7171/jbt.20-3102-001
11. Alberti C, Manzenreither RA, Sowemimo I, Burkard TR, Wang J, Mahofsky K, Ameres SL, Cochella L (2018) Cell-type specific sequencing of microRNAs from complex animal tissues. Nat Methods 15(4):283–289. https://doi.org/10.1038/nmeth.4610
12. Faridani OR, Abdullayev I, Hagemann-Jensen-M, Schell JP, Lanner F, Sandberg R (2016) Single-cell sequencing of the small-RNA transcriptome. Nat Biotechnol 34(12):1264–1266. https://doi.org/10.1038/nbt.3701
13. Hagemann-Jensen M, Abdullayev I, Sandberg R, Faridani OR (2018) Small-seq for single-cell small-RNA sequencing. Nat Protoc 13(10):2407–2424. https://doi.org/10.1038/s41596-018-0049-y
14. Isakova A, Neff N, Quake SR (2021) Single-cell quantification of a broad RNA spectrum reveals unique noncoding patterns associated with cell types and states. Proc Natl Acad Sci U S A 118(51). https://doi.org/10.1073/pnas.2113568118
15. Wang N, Zheng J, Chen Z, Liu Y, Dura B, Kwak M, Xavier-Ferrucio J, Lu YC, Zhang M, Roden C, Cheng J, Krause DS, Ding Y, Fan R, Lu J (2019) Single-cell microRNA-mRNA co-sequencing reveals non-genetic heterogeneity and mechanisms of microRNA regulation. Nat Commun 10(1):95. https://doi.org/10.1038/s41467-018-07981-6
16. Hucker SM, Fehlmann T, Werno C, Weidele K, Luke F, Schlenska-Lange A, Klein CA, Keller A, Kirsch S (2021) Single-cell microRNA sequencing method comparison and application to cell lines and circulating lung tumor cells. Nat Commun 12(1):4316. https://doi.org/10.1038/s41467-021-24611-w

Chapter 13

Immunoblot Analysis from Single Cells Using Milo™ Single-Cell Western Platform

Prashant V. Thakkar

Abstract

In this new era of precision medicine, characterization of single-cell subpopulations to better understand disease etiology is paramount. It is thus an opportune time to explore techniques that allow molecular analysis of single cells and to better understand the basis of pathogenesis of diseases like cancer. Single-cell western blotting is one such method that allows analysis of single cells at the protein level. In contrast to traditional western blotting, which relies heavily on bulk analysis of lysates generated from tissues and is often indicative of the population average, this technique allows analysis of lysates from single-cell subpopulations thereby providing a glimpse into cell heterogeneity. The method entails the use of a chip containing 30 μm thick photoactivated polyacrylamide gel spotted with nearly 6400 microwells. Single cells loaded on the chip are captured in the microwells by passive gravity and are then lysed and electrophoresed using the MILO™ single-cell western platform. This method forgoes the use of transfer of proteins on a PVDF and a nitrocellulose membrane, as performed in traditional western blotting, and all other steps including probing of primary and fluorescent secondary antibodies against the protein of interest are performed directly on the chip. The proteins of interest can then be visualized by scanning a chip with the use of a microarray scanner. The entire procedure can be performed in as less as 4–6 h, and thus this method provides several advantages over traditional western blotting.

Key words Single-cell western blot, Immunoblot, Precision medicine, Tumor heterogeneity, Cell heterogeneity

1 Introduction

In the past decade, we have witnessed huge advances in biomedical research, owing to innovations in genomic technologies, diagnostics, and therapeutic biologics. The plethora of biological information has given us important clues for understanding local and global regulation at various levels of genetic and metabolic hierarchy, thereby boosting applied therapeutics [1–3]. These advances have led to a new era of precision medicine, whereby diseases such as cancer are no longer perceived as a single disease. The traditional "one-size-fits-all" approach to cancer therapy included treating all cancer patients with the same therapy regimen. It is now becoming

Miodrag Gužvić (ed.), *Single Cell Analysis: Methods and Protocols*, Methods in Molecular Biology, vol. 2752,
https://doi.org/10.1007/978-1-0716-3621-3_13,

increasingly clear that the treatment of cancer needs to be customized at an individual level following molecular analysis at the level of single cells [4].

Cellular heterogeneity is an intrinsic characteristic of all multicellular organisms. Both normal and tumor tissues exhibit cellular heterogeneity and different cells within the same tissue can differ in function [5]. These functional differences in various cells are a manifestation of variations in the DNA sequence and/or differences in the expression of RNA and protein. Elucidating these molecular differences is critical to identifying various subpopulations that may play an important role in disease etiology and prognosis. Furthermore, diseases like cancer often show clonal selection that further results in a subpopulation of cells that are either resistant to therapy or show metastatic potential and are responsible for disease progression [6]. Investigation of these cell subpopulations is inherently limited by the fact that the readout of any pooled assay that uses bulk tissue represents a weighted average of that population's cellular constituents [7]. Therefore, it is more important than ever to employ methods that target molecular analyses at a single-cell level. Recent technological innovations now allow one to perform single-cell analysis at the level of DNA, RNA, as well as proteins. Such techniques include capture of CTCs from liquid biopsies with the use of microfluidic devices [6], RNA seq followed by picking cells by Cell Celector [8], single-cell RNA seq using Chromium 10× genomics platform [9], etc.

In the quest for further refining techniques that allow molecular analysis at a single-cell level, a new technique referred to as the single-cell western blot/immunoblot has been developed. This method, as the name suggests, allows analysis of proteins at a single cell level using a modified method based on the traditional western blotting of proteins. This method entails capture of single cells into individual microwells present on a chip coated with photoactivated polyacrylamide (scWest chips from ProteinSimple). These cells are then lysed and electrophoresed directly on the chip using the MILO™ single-cell western blot instrument (available from ProteinSimple) [10]. The separated proteins of interest are then probed for primary antibodies followed by fluorescent secondary antibodies. The proteins of interest can then be visualized as fluorescent spots on the chip as visualized using a microarray scanner. The schematic for the single-cell western blot method-using MILO™ single-cell western platform is presented in Fig. 1. This chapter further discusses the details of the method and hopefully serves as a reference for all new and existing users of this technique.

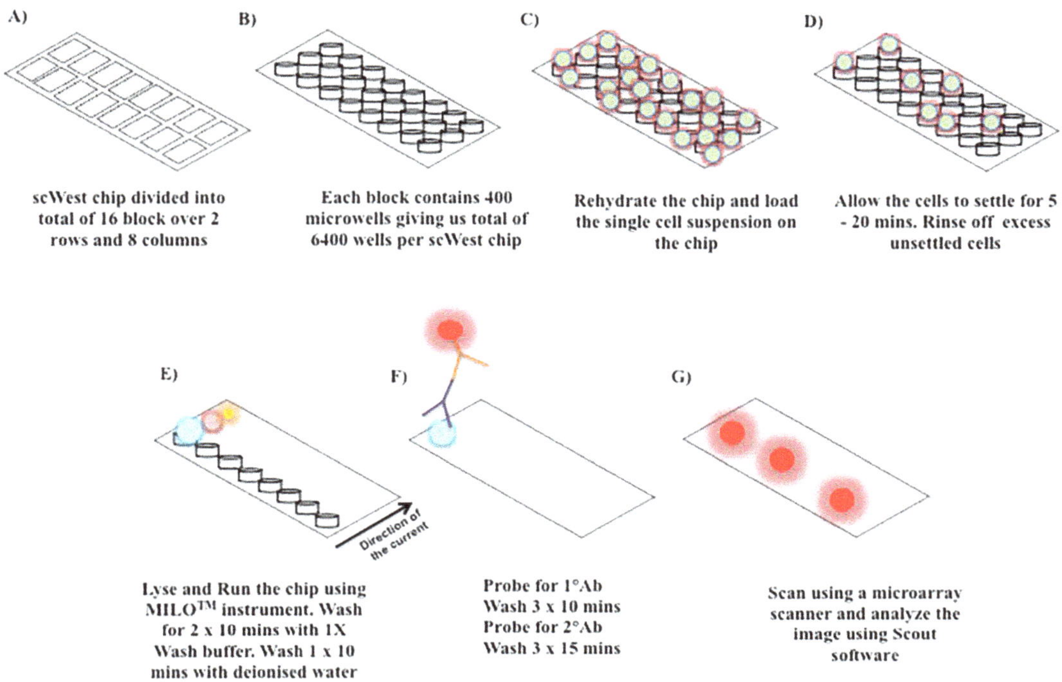

Fig. 1 Schematic of single-cell western blot protocol. Steps **a**–**e** highlight different steps in the single-cell western blot protocol. (**a**) The scWest chip is divided into 16 different grids (displayed over two rows and eight columns). (**b**) Each grid contains 400 microwells thus giving us a total of 6400 microwells per chip. (**c**) The chip is rehydrated with 1× Suspension buffer and a single-cell suspension (approximately 100,000 cells) is loaded on the cells. (**d**) The cells are allowed to settle and nearly 30% of the total microwells are occupied with cells. The excess number of cells is then washed off. (**e**) The cells in the microwells are then lysed and electrophoresed in the same step using the MILO™ single-cell western platform. The chip is washed twice with 1× wash buffer followed by one rinse with deionized water. (**f**) The chip is then probed with primary antibodies against the protein of interest followed by fluorescent secondary antibody. (**g**) The chip is then scanned using a microarray scanner and the resulting image can be analyzed using the Scout software. (Figure adapted from [12])

2 Materials

2.1 Materials Provided with Standard scWest Kit [10]

10× Suspension Buffer.

5× Wash Buffer.

Antibody Diluent 2.

Lysis Buffer.

scWest chips.

2.2 Other Required Materials

Clean 10 cm petri dishes.

Aluminum foil/dark incubation boxes.

Antibody probing chamber (ProteinSimple).

MILO™ single-cell western blot instrument (ProteinSimple).

Metal forceps.

Deionized water/Molecular-grade water.

Primary antibodies.

Fluorescent Secondary antibodies.

Microarray scanner.

Scout software for data analysis (ProteinSimple).

Vortex.

Tabletop centrifuge.

3 Methods

3.1 Preparation of Buffers and Reagents

1. The 10× suspension buffer and 5× wash buffer are provided with the standard scWest kit.
2. Prepare 1× suspension buffer and 1× wash buffer using the above-mentioned 10× and 5× stock solutions respectively using deionized water or molecular grade water [10].

3.2 Priming and Rehydrating scWest Chips

1. Take a scWest chip and place it in the center of a clean 10 cm petri dish with the gel-side facing up. It is important that the dish is clean and sterile since any dust particles that settle on the chip can lead to a high background during the imaging steps. Touching the gel surface of the chip must be avoided since that can add to more dust particles on the chip.
2. Add 10 mL of 1× suspension buffer to the petri dish to completely cover the chip.
3. Incubate the chip for at least 10 min, with slow shaking (<100 rpm) on an orbital shaker before use, so that the chip is completely rehydrated. Alternatively, the chip can also simply be allowed to stand in 1× suspension buffer, although, it is important to make sure that the chip is completely immersed.
4. The chip can be kept in the 1× suspension buffer until the single-cell suspension is ready to be loaded on the chip.

3.3 Preparation of Single-Cell Suspension

1. A single-cell suspension of cell lines can be generated using standard mammalian cell culture procedures.
2. Briefly, remove the cell growth medium and gently wash the dish with 1× PBS (without $MgCl_2$ and $CaCl_2$).
3. Remove 1× PBS and add 2 mL Trypsin solution in a T75 flask (0.05% Trypsin, 0.53 mM EDTA, without sodium bicarbonate) and incubate the flask in a 37 °C incubator for approximately 1–2 min. Incubation times and volume of trypsin can vary depending on the cell lines and size of the flask.

4. Gently tap the flask to detach all the cells and neutralize the trypsin with at least equal volume of growth medium containing 10% FBS.
5. Count cells using hemocytometer or other cell counting equipment and prepare a 5 mL suspension of 10,000 to 100,000 cells/mL in growth medium with 1× suspension buffer.
6. Centrifuge cells at 300 × *g* for 5 min. Remove the growth medium and wash the cells once with 5 mL of 1× PBS (without $MgCl_2$ and $CaCl_2$).
7. Centrifuge cells at 300 × *g* for 5 min. Remove 1× PBS solution and resuspend cells in 5 mL of 1× suspension buffer. The single-cell suspension is now ready for use on the single-cell western blot.

3.4 Loading scWest Chips with Single-Cell Suspension

1. Remove the scWest chip from the petri dish containing 1× suspension buffer and tilt it on one side to remove excess liquid. A paper tissue can be placed underneath to help drain off the excess liquid ensuring that paper tissue does not touch the top of the gel. Place the chip in a new and clean 10 cm petri dish. For selecting appropriate scWest chips, *see* **Note 1**.
2. Load 1 mL (approximately 100,000 cells) of the single-cell suspension dropwise on the scWest chip [10]. Ensure that all areas of the chip are fully covered (Fig. 1c). For using lower number of cells, *see* **Note 2**.
3. Allow the cells to settle for 5–20 min. This step allows the cells to enter and settle in the wells (Fig. 1d). For optimizing appropriate settling time for different cells, *see* **Note 3**.
4. Once the cells are allowed to settle for the desired time, tilt the slide to drain off the liquid and remove the uncaptured cells from the top of the slide. To efficiently remove all the cells from the top of the chip, tilt the petri dish containing the chip at a 45° angle and carefully run the 1× suspension buffer from top to bottom using a 1000 uL pipette.
5. At this step, the chip can be viewed under a bright field microscope to visualize settling of the cells in microwells. Immerse the slide back into the 10 cm petri dish containing 1× suspension buffer. This chip is now ready to be run on MILO.

3.5 Running scWest Chip Using MILO™ Instrument

1. Turn on the MILO instrument and use the touch screen to create optimized user settings. Most typically the conditions listed below can serve as a good reference point [10]. Further optimization may be required for determining optimal conditions.

   Lysis time: 10–15 s.

   Electrophoresis run time: 60–90 s.

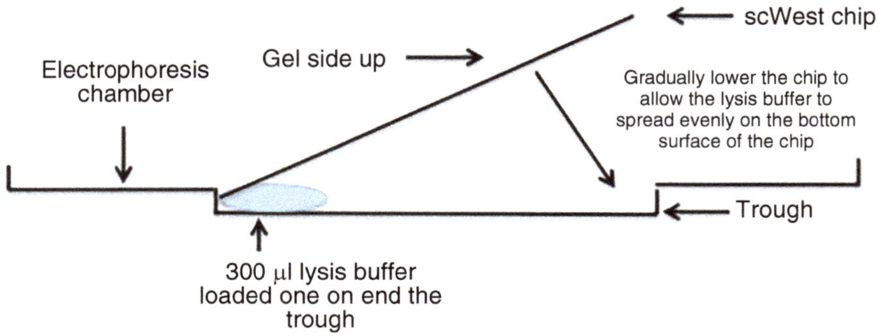

Fig. 2 Methodology to load the chip in the MILO™ instrument to ensure efficient lysis: 300 μL of lysis buffer is loaded on the shorter end of the trough in the electrophoresis. Tilt the chip at a 45° angle and align the shorter end of the chip to the shorter end of the trough where lysis buffer was added. Gradually lay the chip from the other end so that the chip now lays flat and perfectly aligned in the trough. This will cause the lysis buffer to spread throughout the bottom surface of the chip

Electrophoresis voltage: 240 V.

UV exposure time: 4 min.

In our experience, shorter lysis times lead to sharper and more concentrated spots at the visualization step as opposed to more diffused spots seen with longer lysis time.

2. Open the top lid of the MILO instrument and add 300 μL of the lysis buffer provided in the Standard scWest kit along one of the shorter ends of the trough of the white electrophoresis chamber. Do not add more than 300 μL of the lysis buffer since the excess buffer will overflow on top of the gel causing lysis to begin immediately [10].

   The lysis buffer provided with the scWest kits from ProteinSimple does not contain reducing agents. Prior to performing lysis, reducing agents can be added to the lysis buffer. For more information regarding addition of reducing agents, see ***Note 4***.

3. Tilt the chip at a 45° angle and align the shorter end of the chip to the shorter end of the trough where lysis buffer was added in **step 2** above. Gradually lay the chip from the other end so that the chip now lays flat and perfectly aligned in the trough. This will cause the lysis buffer to spread throughout the bottom surface of the chip (Fig. 2).

   This step is important since any bubbles or gaps in the layer of lysis buffer underneath the chip may result in inconsistent electrophoresis of the gel. Handle the chip by only touching it on the side and not the top of the gel. Alternatively, forceps may also be used to avoid physical contact with the gel.

4. Pour the rest of the lysis buffer into the chamber on top of the chip, close the lid, and hit run immediately [10]. It is important

to run the instrument immediately at this step since any delays at this step can cause higher than desired incubation times.

5. Once the run is complete, remove the scWest chip from MILO™ and perform two washes with 1× wash buffer with gentle shaking on an orbital shaker for 10 min each.
6. At the end of the second wash, remove wash buffer and repeat wash with 10 mL of deionized water with gentle shaking on an orbital shaker for 10 min. This step removes the excess salt from the wash buffer and thus will result in low background during the imaging steps.
7. At this step, one can proceed with probing the chip with the primary and secondary antibodies of interest. Alternatively, the chip can be dried overnight and stored at room temperature for months until one is prepared to probe the chip with primary and secondary antibodies.

3.6 Probing scWest Chips with Primary and Fluorescent Secondary Antibodies

1. Dilute the primary antibody to the final concentration of 0.2 μg/μL to 0.05 μg/μL using the "antibody diluent 2" solution provided in the kit [10].

   The protocol is designed to use as little as a total volume of 80 μL of the antibody at a dilution of 1:5 to 1:20, that is, for a 1: 20 dilution add 4 μL of the antibody with 76 μL of the "antibody diluent 2" solution. Most commercially available antibodies should work just fine within this concentration range. This concentration range can serve as a good reference point, although further optimization may be required. Primary antibodies of two different species can also be multiplexed in the same step. Thus, to make an antibody solution of two different primary antibodies at 1:20 dilution each, add 4 μL of each of the antibodies with 72 μL of the antibody diluent. Alternatively, the antibodies against two different proteins of interest, even though from two different species, can also be probed sequentially.
2. A special probing chamber (available through ProteinSimple) [10] is used for antibody incubation. Pipette 80 μL of antibody solution on one end of the probing chamber and place the chip with gel side facing down so that the antibody solution spreads evenly and covers the entire gel surface of the chip. To ensure that the antibody covers the entire gel surface, the chip should be held steadily on one end where the antibody solution was loaded and gradually lowering the chip from the other end (Fig. 3).
3. Avoid bubbles since no antibody signal will be detected in the spot of the bubble. Adjust the chip to eliminate the bubbles if any.
4. Incubate the chip with primary antibody at room temperature for 1 h.

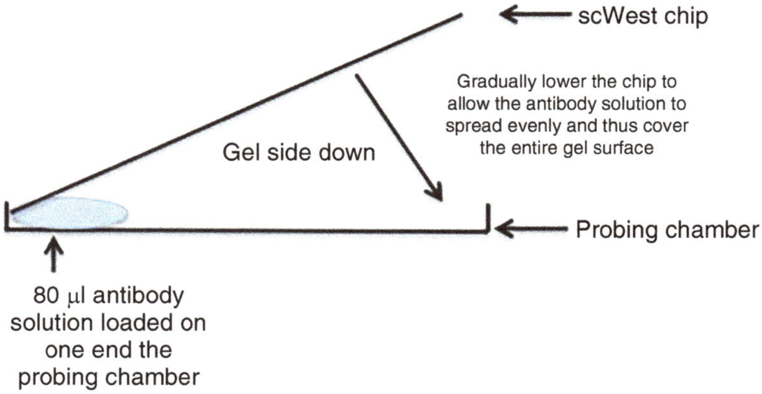

Fig. 3 Methodology to load the chip on antibody probing chamber while probing it with primary and fluorescent secondary antibodies: Pipette 80 μL of secondary antibody solution on one end of the probing chamber and chip is placed with gel side facing down so that the antibody solution spreads evenly and covers the entire gel surface of the chip. The chip should be held steadily on one end where the antibody solution was loaded and gradually lowering the chip from the other end

5. At the end of 1-h incubation, carefully lift the chip from one end while holding the other end so as not to tear the gel. Invert it with the gel side facing up in a 10 cm petri dish.
6. Wash it three times with 1× wash buffer, while gently shaking on an orbital shaker for 10 min each.
7. At the end of the third wash, prepare the fluorescent secondary antibody solution using the "Antibody diluent" at 1:20 to 1:40 dilution [10].

   Similar to primary antibody incubation, secondary antibodies against primary antibodies from two different species can be multiplexed. It is important that the two secondary antibodies against the two different primary antibodies fluoresce in different spectral channels to ensure specificity of the signal.
8. For secondary antibody incubation, pipette 80 μL of secondary antibody solution on one end of the probing chamber and place the chip with gel side facing down so that the antibody solution spreads evenly and covers the entire gel surface of the chip. To ensure that the antibody covers the entire gel surface, the chip should be lowered slowly keeping one end steadily rested on the probing chamber and gradually lowering the chip from the other end (Fig. 3).
9. Incubate the secondary antibody solution for 1 h, this time covering it with either a black box or aluminum foil so as to protect it from light.

10. At the end of the 1-h incubation, carefully lift the chip from one end while holding the other end so as not to tear the gel. Invert it to with the gel side facing up in a 10 cm petri dish.
11. Wash it three times with 1× wash buffer, while gently shaking on an orbital shaker for 15 min each this time also protecting from light [10].
12. After the last wash, proceed to incubating primary antibody for the next protein interest. If all the proteins of interest have been probed with their respective primary and secondary antibodies, air dry the chip in the dark for about 40–45 min.
13. The chip is now ready for imaging.

3.7 Imaging of scWest Chips

1. scWest chip can be scanned using many of the recommended fluorescence microarray scanners recommended by ProteinSimple [10]. In case where the microarray scanner is not available, *see* **Note 5** for the use of Zeiss spinning disc microscope for scanning images.
2. We have personally used GenePix4000B from Molecular Devices [11], which in our experience is very user-friendly to use. However, this instrument can only allow for scanning of up to two different fluorescent channels (excitation wavelengths of 532 nm and 635 nm with emission wavelengths in greenish yellow to red/far red regions of the spectrum) whereas other microarray scanners can allow up to four or even five different channels thus allowing for probing of more proteins per chip.
3. To start scanning the slide using the GenePix4000B, open the door and lift the slide holder. Push the slide clamp to place the slide in its position with gel side facing down. Lower the slide holder back to its original position and close the door.
4. Close the instrument door and start the "Genepix pro image analysis software" on the computer attached to the instrument.
5. Once the software interface is open, click on "Hardware settings" to select the wavelength the chip would be scanned at. For each wavelength, set the scan pixel resolution to 5 μm and adjust the PMT gain to maximize the brightness of the signal without saturating the detector. Change the power of each laser to 100%.
6. Once all the settings are set, one can hit preview scan to get a quick preview of image the scan would generate. If the preview scan looks good, one can then hit "Data Scan" for scanning the chip using both wavelengths of "Single wavelength scan" option to scan the chip only using one wavelength.
7. The instrument allows simultaneous scans at two wavelengths and scans the standard scWest chip (approximately

25 mm × 75 mm) at a resolution of 5 μm in approximately 12 min. The resulting image is saved as a 16-bit TIFF file, which can be easily exported into the Scout software for further analysis.

3.8 Data Analysis Using Scout Software

1. The images generated using microarray scanners for any given chip are saved as separate file per channel. These images are saved in .tiff format, and this format can be easily used for analysis using Scout software. Export/Add all the images for a given chip from "File" menu in the Scout software [10].
2. The software requires that each exported image be registered. To do this, select the first block of the first row and the eight block in the second row (i.e., two diagonally opposite grids). Then click on the first well in the first block and the last well in the last block. This step helps ensure that the image alignment is correct and any misalignment can be corrected [10].
3. Following **step 2**, the software will then automatically detect the lanes with a positive signal/peaks. This detection can be then manually inspected, and these parameters can be adjusted to define the true signal under "Scan properties" table [10].
4. Once all the peaks are detected, go to the peak table to identify and name the peak of interest as the appropriate molecular weight size peak.
5. The software can then provide statistics such as total number of microwells occupied, total number of wells that are positive in each spectral channel, etc. A representative image of the results obtained from this single-cell western blot method is presented in Fig. 4.

3.9 Limitations

Although this technique shows incredible promise and is huge advancement in allowing molecular analysis of proteins at the level of single cells, it is not without limitations. Some of these limitations are highlighted below:

1. The method currently does not incorporate the use of molecular weight markers that can be run simultaneously with the sample. This limitation then requires one to probe for one of the housekeeping genes such as Tubulin or GAPDH to be used as a size marker. One can then predict the size of the protein of interest based on the size and migration of the housekeeping gene. In addition, since one has to always probe for a housekeeping gene, it expends one of the fluorescent channels, leaving one less channel that can be used for probing an additional protein of interest.
2. Although migration of protein occurs over a short distance of 900 μm, and while the migration of proteins of different molecular sizes follows a linear range in that short distance,

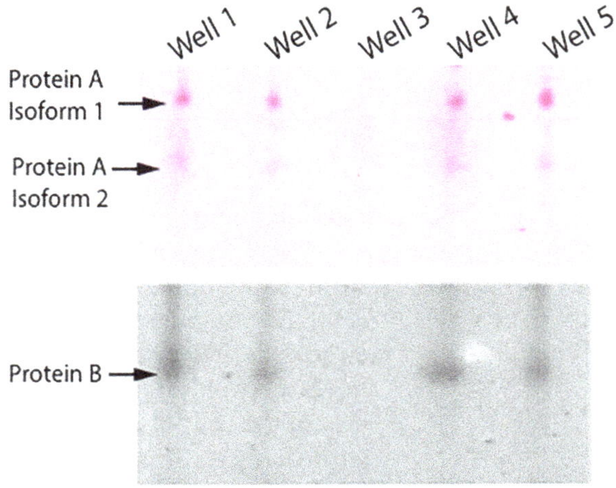

Fig. 4 Representative image of the result obtained from a single-cell western blot: Single-cell western blot protocol was performed using the Standard scWest chip and was probed for two antibodies against protein A and protein B. Two isoforms that differ in size by 30 kDa can be visualized in the top panel and the housekeeping gene (Protein B) can be visualized in the bottom panel

deducing molecular weights of proteins of interest in reference to the housekeeping gene can sometimes still be inconclusive. In particular, some of the proteins migrate as dimers and multimers since the migration of proteins occurs under non-reducing, non-denaturing conditions. For example: Proteins such as cytokeratin and GAPDH, which would normally migrate as 67 kDa and 38 kDa monomers respectively on a western blot, these proteins migrate as 130 kDa dimer and approximately.

3. Currently, the system only allows for analysis of proteins in the molecular weight size range from 175 kDa to 15 kDa. Hence, proteins higher than 175 kDa cannot be studied using this system and will need to be studied by traditional western blotting. Furthermore, the scWest chips available are only available in single gel composition of 8% polyacrylamide. As of now, no gradient gels are available, and hence proteins with molecular sizes at the extreme ends of the above-mentioned range can only be resolved with short run times, which then would compromise on the resolution of proteins in the same spectral channel. Even while resolving proteins similar in size, the system requires that the proteins differ in size by at least 10% if viewed in different spectral channels and at least 30% when viewed in the same spectral channel. For example: two isoforms of the same protein, and thus viewed in the same spectral channel, cannot be resolved on the same scWest chip

unless they differ from each other in molecular weight size by at least 30%.

4 Notes

1. *Choosing appropriate scWest Chips for the cells of interest*

   There are three different types of scWest chips available from ProteinSimple, namely (1) Small scWest kit for small cells, (2) Standard scWest kit, and (3) Large scWest kit for large cells. These kits differ in the diameter of microwells, and hence appropriate kit must be chosen depending on the diameter of the cells of interest. Small scWest kit can be used for cells with diameter less than 10 μm, standard kits can be used for cells with diameter in the range of 10–20 μm, and large kits can be used for cells with 20–30 μm [10].

2. *Using the appropriate number of cells*

   The scWest chips are divided into a total of 16 grids over two rows and eight columns. Each block has 400 microwells divided over 10 rows of 40 microwells each. This layout gives a total of 6400 total microwells per scWest chip. When cells are loaded on the chip, the cells settle into the microwells by passive gravity. This process follows a Gaussian distribution where only 20–30% of the microwells are occupied, with nearly 10–15% of the occupied microwells contain single cells and the remaining occupied microwells contain two or three cells, that is, doublets or triplets [10]. This percentage occupancy is even lower with lower number of cells. In situations where the starting material is not a limiting factor, for example, cell lines, 100,000 cells should be loaded on the chip. The recommended minimum number of cells to be loaded on the chip is 10,000 cells in 1 mL volume over the entire chip. We have tried loading as little as 1000 cells and were able to capture only 15–20 cells on the chip when loaded in 1 mL of suspension buffer. Alternatively, loading a lower number of cells in as little as 100–200 μL of suspension buffer can help improve the capture efficiency of cells. This is especially important in cases where a specific subset of cells, isolated using sorting by a flow cytometer or concentrated by means of specific cell surface markers, needs to be analyzed, and thus the starting material is limited.

3. *Optimizing settling time for the cells on scWest chip*

   Since settling of the cells into microwells is driven by passive gravity, 10 min incubation leads to capture of 0–4 cells/microwells and occupancy of approximately 30% of the microwells. Longer incubation times result in capture of higher number of cells but also lead to a higher percentage of microwells with more than one cell per microwell, that is, doublets and triplets per microwell [10].

4. *Use of reducing agents to completely denature proteins*

The lysis buffer provided in the kit currently does not contain any reducing agents. Furthermore, since the lysis step is performed directly on the chip once cells are loaded in the microwells, there is no boiling step, which may result in incomplete denaturation of proteins as well as incomplete dissociation of oligomers. Reducing agents can be easily incorporated in the protocol by using 1× SDS buffer for lysis of cells in the microwells. Although this buffer is highly ionic, the conductance of 1× SDS buffer is higher compared to the lysis buffer provided in the kit. Hence, the voltage should be reduced from 240 V to 150 V and run time should be increased by approximately 50–60%. Alternatively, 10 mM TCEP (tris-(2-carboxyethyl)phosphine) (pH 7.0) can be added to the lysis buffer provided to the kit. This is important since the increased conductance by the lysis buffer can lead to heating up of the gel, which may in turn affect the resolution.

5. *Use of Zeiss spinning disc microscope for scanning scWest chips*

In cases where microarray scanners are hard to find or are not available in one's institution, we have also successfully employed Zeiss spinning disc microscope where one can scan the entire chip in "tiles" and then "stitch" all the images to create a single image of the entire chip. This image can then be exported to the Scout software for further analysis.

References

1. Ashley EA, Butte AJ, Wheeler MT, Chen R, Klein TE, Dewey FE, Dudley JT, Ormond KE, Pavlovic A, Morgan AA, Pushkarev D, Neff NF, Hudgins L, Gong L, Hodges LM, Berlin DS, Thorn CF, Sangkuhl K, Hebert JM, Woon M, Sagreiya H, Whaley R, Knowles JW, Chou MF, Thakuria JV, Rosenbaum AM, Zaranek AW, Church GM, Greely HT, Quake SR, Altman RB (2010) Clinical assessment incorporating a personal genome. Lancet 375(9725):1525–1535. https://doi.org/10.1016/S0140-6736(10)60452-7
2. Choi M, Scholl UI, Ji W, Liu T, Tikhonova IR, Zumbo P, Nayir A, Bakkaloglu A, Ozen S, Sanjad S, Nelson-Williams C, Farhi A, Mane S, Lifton RP (2009) Genetic diagnosis by whole exome capture and massively parallel DNA sequencing. Proc Natl Acad Sci U S A 106(45):19096–19101. https://doi.org/10.1073/pnas.0910672106
3. Ng SB, Turner EH, Robertson PD, Flygare SD, Bigham AW, Lee C, Shaffer T, Wong M, Bhattacharjee A, Eichler EE, Bamshad M, Nickerson DA, Shendure J (2009) Targeted capture and massively parallel sequencing of 12 human exomes. Nature 461(7261):272–276. https://doi.org/10.1038/nature08250
4. Ashley EA (2016) Towards precision medicine. Nature reviews. Genetics 17(9):507–522. https://doi.org/10.1038/nrg.2016.86
5. McGranahan N, Swanton C (2017) Clonal heterogeneity and tumor evolution: past, present, and the future. Cell 168(4):613–628. https://doi.org/10.1016/j.cell.2017.01.018
6. Ignatiadis M, Dawson SJ (2014) Circulating tumor cells and circulating tumor DNA for precision medicine: dream or reality? Ann Oncol 25(12):2304–2313. https://doi.org/10.1093/annonc/mdu480
7. Yuan GC, Cai L, Elowitz M, Enver T, Fan G, Guo G, Irizarry R, Kharchenko P, Kim J, Orkin S, Quackenbush J, Saadatpour A, Schroeder T, Shivdasani R, Tirosh I (2017)

Challenges and emerging directions in single-cell analysis. Genome Biol 18(1):84. https://doi.org/10.1186/s13059-017-1218-y

8. Zhu S, Qing T, Zheng Y, Jin L, Shi L (2017) Advances in single-cell RNA sequencing and its applications in cancer research. Oncotarget 8(32):53763–53779. https://doi.org/10.18632/oncotarget.17893
9. Haque A, Engel J, Teichmann SA, Lonnberg T (2017) A practical guide to single-cell RNA--sequencing for biomedical research and clinical applications. Genome Med 9(1):75. https://doi.org/10.1186/s13073-017-0467-4
10. MILO™ user guide from ProteinSimple. https://www.proteinsimple.com/documents/Milo_User_Guide_R5.pdf
11. Genepix4000B Microarray scanner user guide. http://mdc.custhelp.com/euf/assets/content/GenePix_4000B_UserGuide.pdf
12. Hughes AJ, Spelke DP, Xu Z, Kang CC, Schaffer DV, Herr AE (2014) Single-cell western blotting. Nat Methods 11(7):749–755. https://doi.org/10.1038/nmeth.2992

Chapter 14

Imaging of Subcellular Distribution of Platinum in Single Cells Using Laser Ablation Inductively Coupled Plasma Mass Spectrometry

Amy J. Managh and Calum J. Greenhalgh

Abstract

Laser ablation inductively coupled plasma-mass spectrometry (LA-ICP-MS) is a well-established and sensitive analytical technique, which provides high-resolution imaging of endogenous elements, element tagged-markers, metal-containing nanoparticles, and metallodrugs within cells. Here we describe a protocol for imaging the subcellular distribution of platinum within A549 cells, following their incubation with the platinum-based anticancer agent, Oxaliplatin. We outline the essential steps in sample preparation and instrumental setup and discuss how the current generation of low-dispersion instruments facilitates new approaches to data acquisition and image processing. The protocol described herein can be easily adapted for other cell lines and metal-containing labeling agents.

Key words Laser ablation, ICP-MS, Imaging, Oxaliplatin, A549 cells

1 Introduction

Laser ablation inductively coupled plasma-mass spectrometry (LA-ICP-MS) is a sensitive elemental analysis technique, which enables highly specific sampling and imaging of solid materials with micrometer scale resolution. Although traditionally used for geological analysis, the technique is now well-established for a range of biological applications. Examples include visualization of drugs and tagged markers in tissue sections [1, 2], as well as quantitative measurements of endogenous elements and isotope ratio distributions [3, 4]. Due to its high spatial resolution, the technique is particularly suited to cellular and subcellular imaging. The first use of LA-ICP-MS for subcellular histology was reported in 2011 [5]. Geisen and co-workers used a 4 μm diameter laser beam to differentiate the nucleus and cytoplasmic regions of iodinated fibroblast cells. Subsequent studies have reported the use of lanthanide-tagged complexes and iridium intercalators to visualize

Miodrag Gužvić (ed.), *Single Cell Analysis: Methods and Protocols*, Methods in Molecular Biology, vol. 2752,
https://doi.org/10.1007/978-1-0716-3621-3_14,

the protein distribution, DNA content, cell volume, and cell cycle phase of individual cells [6, 7]. LA-ICP-MS has also been used to determine nanoparticle and drug uptake by cells. Studies have reported both quantitative determination of Au nanoparticle uptake by whole cells [8] and subcellular imaging of their localization to different cellular compartments [9]. Furthermore, the technique may be vital for monitoring the fate of rare cell populations used in cell-based treatments. In a 2013 study, individual regulatory T cells that were labeled with the MRI contrast agent Omniscan were successfully identified in mouse peritoneal lavage samples at 10 days post-administration [10]. Subsequent publications have also reported LA-ICP-MS identification of individual therapeutic cells within ex vivo tissue sections in similar cell tracking experiments [11, 12].

Despite the success of the above applications, LA-ICP-MS has not yet gained full traction within the molecular biology community. This is mainly due to inefficiencies associated with the previous generation of instruments, in particular in the interface between the LA and ICP-MS units, which resulted in broadening of signals and hence lengthy analysis times compared to conventional imaging methods. In recent years a number of new interfaces have been developed to combat this problem [13–15]. These new technologies provide vastly improved aerosol transport characteristics, leading to faster signal responses, higher absolute sensitivity, and improved signal-to-noise ratios. Importantly, these new developments are now starting to reach the market. In this protocol we describe a method for qualitative single-cell analysis and imaging of platinum distribution in Oxaliplatin spiked A549 cells, using one of these new laser ablation platforms. We focus on how the improved aerosol transport characteristics have created an opportunity to move away from the traditional "rastered" method of imaging, toward an approach that preserves maximum image resolution for individual cell analysis.

2 Materials

2.1 Equipment

1. Element XR sector-field ICP-MS (Thermo Fisher Scientific).
2. NWR image laser ablation system (Elemental Scientific Lasers).
3. Hettich Universal 320R centrifuge, fitted with 4 place rotor with fixed angle chambers.
4. Laminar flow cabinet.
5. Microscope.
6. Hemocytometer.

2.2 Reagents

1. A549 cell line.
2. RPMI-1640 medium, containing L-glutamine and sodium bicarbonate, supplemented with 10% Fetal Bovine Serum (FBS).
3. Dulbecco's modified phosphate-buffered saline (PBS).
4. Trypsin/EDTA.
5. Trypan blue.
6. High purity (99.996%) helium and argon supply.

2.3 Other Consumables

1. T-25 cell culture flasks.
2. Glass microscope slides (*see* **Note 1**).
3. Angle cyto chamber (DJB Labcare).
4. Filter cards for 1, 2, and 4 mL Cyto chambers (DJB Labcare).
5. Conical tubes (15 mL; polypropylene or polystyrene, e.g., from Falcon).

3 Methods

3.1 Preparation of Cells for Analysis

Cell culture and labeling conditions will vary depending on the nature of the cell line being studied. The instructions below are based on the culture of A549 cells, which are an adherent human lung adenocarcinoma cell line, and their labeling with the anticancer drug, Oxaliplatin. For other cell lines and labeling approaches, you should follow your usual procedures for that cell line, and then proceed to **step 9**.

1. Seed cells at an initial density of 1×10^6 cells in 5 mL of complete media. Incubate at 37 °C in 5% CO_2 overnight.
2. Add 50 μM Oxaliplatin to the culture media and leave to incubate for a further 24 h.
3. Aspirate the used media. Add 5 mL PBS to the flask (*see* **Note 2**) and gently swirl for 1 min ensuring the entire surface is covered. Aspirate the PBS into a waste container. Repeat 2–4 times (*see* **Note 3**).
4. Add 1 mL of warm trypsin-EDTA solution to the flask and incubate at 37 °C for 3 min.
5. Inspect the flask under a microscope to check that the cells have successfully detached from the surface (*see* **Note 4**), then add 4 mL of complete media to quench the trypsin.
6. Set aside a 100 μL aliquot of the cell suspension for cell counting.
7. Transfer the remaining cells into a 15 mL falcon tube and centrifuge at 400 rpm for 5 min.

8. Mix the 100 μL aliquot of the cell suspension with an equal volume of 0.4% trypan blue solution. Dispense 10 μL of this solution into a hemocytometer and count the cells (*see* **Note 5**).
9. Remove the supernatant from the centrifuged cells (*see* **Note 6**) and dilute the cell pellet in phosphate-buffered saline to a concentration of 100,000 cells/mL. Mix the solution well.
10. Disperse the cells onto a glass microscope slide using a Cyto-centrifuge (*see* **Note 7**). Place a filter card onto a clean glass microscope slide, then clamp an angled cyto chamber in place over the filter card, using the locking clip provided. Place the cyto chamber into the fixed-angled chambers in the centrifuge, with the funnel inlet facing upwards. Transfer 0.5 mL of the diluted cells to the funnel. Ensure that the centrifuge is balanced, then centrifuge at 35 rcf for 3 min to disperse the cells across the slide. Allow the slide to air-dry in a dust-free environment for half an hour before performing the analysis. If necessary the slides can be stored for several weeks in a cool, dry, and dust-free environment.

3.2 Setting Up the LA-ICP-MS System

1. Insert the slide into the laser ablation chamber and secure it in place (*see* **Note 8**). Ensure that the chamber is appropriately sealed afterwards by fastening the screws on the outside of the chamber (finger-tight).
2. Connect the inlet of the laser ablation chamber to a helium supply and connect the outlet of the chamber to the ICP-MS torch, using suitable tubing, e.g., Tygon. Allow for the introduction of an argon make-up flow between the ablation chamber and the ICP torch (*see* **Note 9**).
3. If available, connect a trigger cable between *SyncOut* on the laser ablation system and the external trigger in port on the mass spectrometer. This is to synchronize the firing of the laser with the acquisition cycle of the mass spectrometer.
4. Purge the chamber with helium for around 10 min by selecting the *purge* button in the laser software. While the instrument is purging it is useful to move the inner cup around the chamber to aid removal of residual air from the corners of the chamber. This can be done using the x,y controls.
5. Light the plasma (*see* **Note 10**). Once the plasma has lit set the helium flow through the laser ablation chamber to around 1.4 L/min. It is useful to do this at a low ramp rate of around 20 mL/min in order to avoid extinguishing the plasma. Allow the plasma conditions to stabilize for approximately an hour before performing any analysis.
6. A sample to cup distance of 0.2–0.3 mm is optimal for efficient material transport out of the chamber. To set this use the z

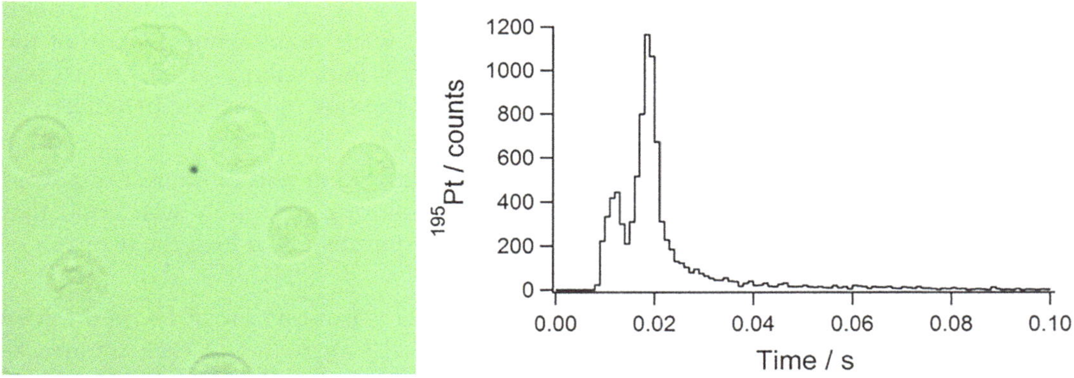

Fig. 1 Left image: Microscopic image showing A549 cells that were dosed with 50 μM Oxaliplatin. The Cytospun cells are spaced an adequate distance for single-cell targeting. Right: LA-ICP-MS signal for the analysis of the Pt in the cell on the far left. Data points in raw counts (shown as horizontal lines) are spaced at 1 ms intervals allowing the full peak profile to be visualized

controls within the laser software to focus on the lower inner edge of the sample cup, and then focus down a further 0.2–0.3 mm. Use the adjustment screws on the sample drawer to bring the glass slide up into focus (*see* **Note 11**).

7. Use the x,y controls to move to the location of interest on the slide. You may need to refocus in the z direction afterwards if the slide is not perfectly level. When in focus the cells should appear similar in nature to the image in Fig. 1. If necessary, re-center the cup so that the crosshairs on the screen are positioned in the center of the aperture in the cup.
8. Warm up the laser before performing any analysis by firing the laser for a few minutes at a high energy and repetition rate, with the shutter closed. The *Emission* button can also be used to reinitialize the laser after periods of downtime.
9. If the sample chamber needs to be opened while the plasma is lit, use the bypass button to avoid entrainment of air. Once closed, the chamber will need to be re-purged before proceeding to *online* mode (*see* **Note 12**).

3.3 LA-ICP-MS Analysis of Pt Content in Whole Cells

Here, analyzing the platinum content of whole cells is used to qualitatively assess the heterogeneity in Oxaliplatin uptake across the cell population.

1. In the LA software select the spot tool, and then left-click on the center of one of the cells of interest.
2. Right-click the above spot and amend its properties. To ensure the entire cell is ablated, set a spot size that is at least 5–10 μm larger than the diameter of cells of interest (for this analysis a spot size of 60 μm was used). Set the repetition frequency as

1 Hz. Select a laser energy that is sufficient to ablate through the entire depth of the cell, while minimizing ablation of the glass substrate (in this work a laser energy of ~5 J/cm^2, was used—*see* **Note 13**). Check the box to set these parameters as default.

3. Click on the center of the remaining cells of interest. To avoid accidental ablation of neighboring cells, only select cells that are separated by a distance equivalent to at least the diameter of the laser beam.
4. In the LA software, set a laser warm-up time of 1 s. In the *Run Experiment* window, select *All Patterns* and click the checkboxes *Manually Control Valves* and *Enable Laser During Scans.* In the laser *SyncOut* options, set the trigger to *Active During Pattern Scan.*
5. Ensure that the ICP-MS is in speed or time-resolved mode and set up an appropriate ICP-MS method and sequence. First, select the mass values to be detected (in this study the ^{195}Pt isotope was measured). Ensure that the segment durations of the method (i.e., the data intervals of the output signal) allow more than one data point per ablation event (*see* **Note 14**). Determine the analysis time for the selected patterns and ensure that the length of the ICP-MS method is equivalent to this duration, plus an extra 5–10 s.
6. Click *Run* in the ICP-MS software and wait until the "awaiting external trigger" message appears in the sequence window.
7. Click *Run* in the laser software.
8. Export the data as a time-resolved data file. Plot the graphs in Excel, then find and integrate the area under each of the peaks (*see* **Note 15**).

3.4 LA-ICP-MS Imaging of Pt Distribution Across Cells

Here, subcellular imaging is used to qualitatively assess the distribution of platinum within individual cells, in order to assess whether the drug is reaching its target site of action (binding to DNA in the nucleus).

1. In the LA software, use the line tool to draw a single horizontal or vertical line that is long enough to span the region of interest at its widest point. If necessary, this line can be subsequently resized or repositioned to the edge of the region of interest, using the *Edit Endpoints* function.
2. Right-click the above line and amend its properties. Choose a laser beam diameter that is smaller than the dimensions of the cells of interest (in this work a 2 μm spot size was used). Set the highest possible repetition frequency that enables baseline resolution between the peaks from consecutive ablation events (in this work 20 Hz was used). Select a scan rate that is

equivalent to the spot size multiplied by the repetition frequency of the laser (e.g., for the settings above, a scan speed of 40 μm/s was selected). These settings avoid "over-sampling" and preserve image resolution by ensuring that the data from adjacent, non-overlapping locations is resolved in the time-domain. Finally, set a laser energy that is sufficient to ablate through the entire depth of the cell, while minimizing ablation of the glass substrate (in this work a laser energy of ~4 J/cm^2 was used—*see* **Note 13**).

3. Duplicate the above line an appropriate number of times to cover the region of interest. Ensure that these duplicate lines are offset by a space equivalent to the size of the laser beam.
4. Ensure that the ICP-MS is in speed or time-resolved mode and set up an appropriate ICP-MS method. First, select the mass values to be detected (in this study the ^{195}Pt isotope was measured). Ensure that the segment durations of the method (i.e., the data intervals of the output signal) are compatible with the repetition rate of the laser (*see* **Note 14**). Determine the analysis time for a single ablated line (the length of the line divided by the laser scan speed). The length of the ICP-MS method should be equal to this time, plus an extra 1–2 s.
5. Set up a sequence in the ICP-MS software. Ensure that the number of samples in the ICP-MS sequence is equal to the number of lines in the laser method. Set the method to trigger the acquisition when a signal from the laser system is received. In the Element XR software the relevant settings file for this (*ext_trigger.inl*) can be found under *Sampling* in the *Sequence* window. Other instruments will have a similar option.
6. In the LA software, set a laser warm-up time of 5 s and a washout time of 5 s. The washout time is a buffer to ensure that the ICP-MS is waiting to receive the trigger signal when ablation of the next line begins. In the *Run Experiment* window, select *All Patterns* and click the checkboxes *Manually Control Valves* and *Enable Laser During Scans*. In the laser *SyncOut* options, set the trigger to *Active During Pattern Scan* (*see* **Note 16**).
7. Click *Run* in the ICP-MS software and wait until the "awaiting external trigger" message appears in the sequence window.
8. Click *Run* in the laser software. It is advisable to stay with the instrument for the first few ablated lines in order to check that the ICP-MS is triggering correctly.

3.5 Generation of LA-ICP-MS Images

1. The data output from the imaging experiment will consist of a series of files, each file containing the data for one ablated line. Assuming sufficient instrument sensitivity, the data will consist of peaks separated by background (or close to background)

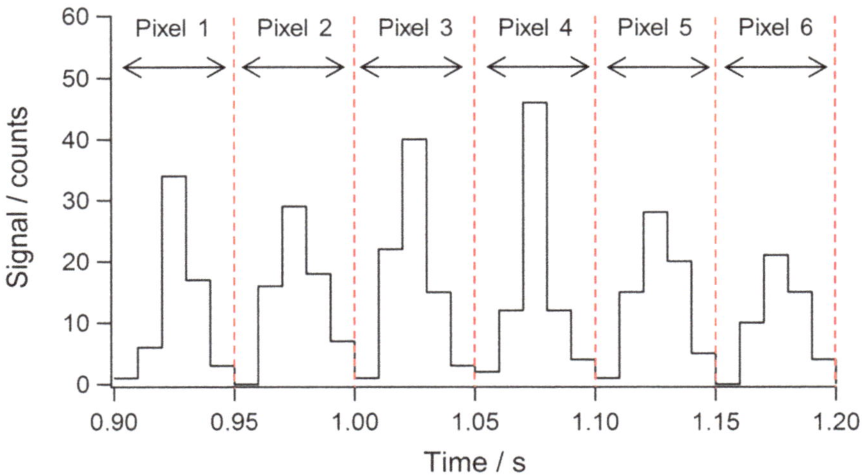

Fig. 2 Schematic depiction of the data processing approach required for a pixel-by-pixel imaging experiment. The instrumental conditions are set such that every laser shot samples an adjacent, non-overlapping area and the signal is allowed to wash out before the next location is sampled. In this example, data points in raw counts (shown as horizontal lines) are spaced at 10 ms intervals in the ICP-MS method, allowing 5 data points to be collected per laser shot (laser at 20 Hz). Markers are then placed between each laser shot and the data between the markers is summed to give a value for each pixel in the final image

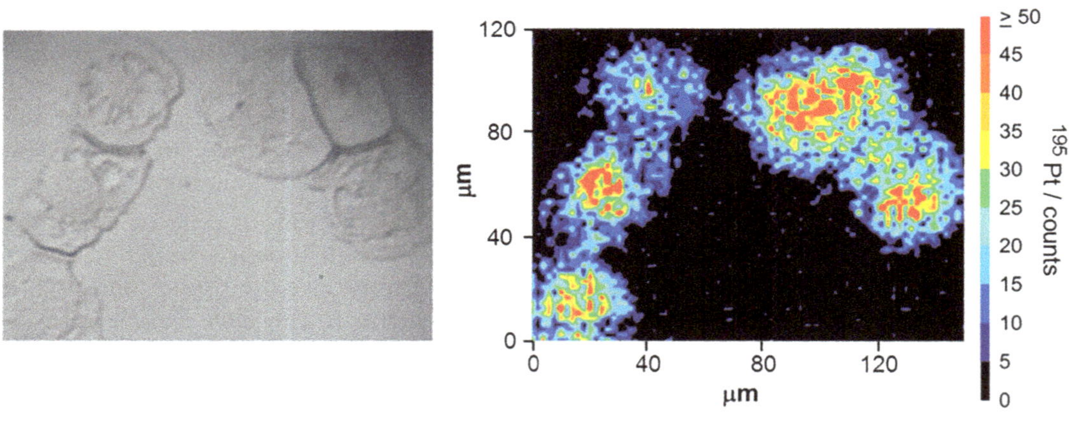

Fig. 3 Left image: Microscopic image of a clump of A549 cells, which were dosed with 50 μM Oxaliplatin. Right: LA-ICP-MS image showing the distribution of Pt across the cells. Cells were imaged using a 2 μm laser beam, with a repetition frequency of 20 Hz and a scan speed of 40 μm/s. It appears that Pt is preferentially accumulated in the nucleus of these cells

signal. An example of the typical output signal is shown in Fig. 2. First, check the data in the acquired files to get a feel for the magnitude of the signals in your data.

2. Sum the signal for each ablated shot, as shown in Fig. 2, to get a list of pixel intensities for each ablated line. Compile these values into a 2D matrix, such that the columns in the matrix

correspond to the ablated lines (placed in order of ablation). This processing can be manually performed within Excel (extremely time consuming) or automated using a custom Excel macro or Matlab script. Alternatively we use a custom app, the LA-ICP-MS Image Tool, to find the peaks in our LA-ICP-MS data, position markers between the peaks and sum the signals between the markers. The app then generates a preview of the final image as a 2D color map (*see* Fig. 3). The app is freely available, has a detailed instruction manual [16], and can process the raw data into images in less than 5 min.

3. Finally, if higher quality images are required, you may wish to export the data to a graphics package for further processing.

4 Notes

1. We recommend Polysine-coated microscope slides (Thermo Fisher Scientific, UK). The surface coating minimizes background signal by providing a barrier between the sample and glass surface.
2. Take care not to touch the cell surface. We recommend adding the liquid down the side of the flask, before gently tipping it onto the cells.
3. Washing the cells several times is required to remove unbound label from the surface of the cells. A higher number of washes may be required depending on the nature of the labeling agent and the concentration of it used.
4. Cells should appear round when detached. Tap the flask to dislodge any unbound cells. If the cells have not been trypsinized, return the flask to the incubator for a further 1 min.
5. The initial cell concentration per mL can be calculated by multiplying the average viable cell count per square by the trypan blue dilution factor (2) and then dividing by the volume (in mL) of the square.
6. Check that a pellet has formed at the bottom of the tube. Take care not to disturb the cell pellet when removing the supernatant.
7. Centrifuging the cells disperses the cells across the slide and helps to remove traces of PBS and culture media, which are absorbed into a filter card. The aim of this step is to obtain sufficient spatial separation between cells to enable single-cell targeting (Fig. 1).
8. It is advisable to fix the slide securely in place inside the chamber to prevent it from moving during the analysis. The NWR image chamber is equipped with a suitable locking clip.

However, for chambers without this, a piece of double-sided sticky tape placed on the back of the slide is an effective substitute to temporarily fix the slide in place. Before fastening down the slide we find that drawing the outline of the cell containing region on the rear side of the slide is helpful in order to help us to move quickly to the location of the cells once the slide is in the chamber.

9. The argon make-up flow can be introduced through a simple T-piece located in the transfer line between the ablation chamber and the ICP torch. However, we prefer to use a Dual Concentric Injector (Elemental Scientific Lasers, USA) to minimize aerosol dispersion and provide improved sensitivity. Regardless of the mechanism of argon introduction, minimizing the total path length between the ablation chamber and the ICP torch is recommended.
10. ICP-MS parameters will vary depending on the ICP-MS instrument used. We find that the optimum Ar sample gas flow is usually much lower with the NWR image system, compared to our normal ICP-MS operation. A sample gas flow of around 0.5 L/min and an RF power of 1100 W are usually optimum for our instrument.
11. If your sample tray does not enable sample height adjustment, skip setting the sample-to-cup distance and use the z controls to focus on the sample instead.
12. The Dual Concentric Injector bypasses the valve system on the laser unit. If using the DCI, it is advisable to clamp the transport tubing with a temporary clip if opening the chamber or purging the system while the plasma is lit.
13. The degree of laser coupling with the glass slide will vary depending on the wavelength of the laser system used. We have found that the 266 nm wavelength is efficient at ablating through biological material, but produces negligible ablation of the glass at the energies described above. For other common wavelengths, e.g., 213 nm and 193 nm, the energy may need to be tuned more extensively due to more efficient coupling of the deep-UV wavelengths with the glass matrix.
14. Ideally the temporal resolution of the ICP-MS method will be such that there is more than one data point per cell event. For example in the peak in Fig. 1 (right hand image) data points are spaced at 1 ms intervals resulting in around 50 data points over the ~50 ms event. Note that the data acquisition speed of some of the older ICP-MS systems is not always sufficient to enable the required resolution for high-speed LA signals. This issue has largely been rectified in the current generation instruments, but plug-in detector boards are a solution to improve the temporal resolution of older systems [17].

15. The peaks can be manually located within the Excel data or this can be automated through the use of in-house Excel macros or Matlab scripts. We use a custom app, the LA-ICP-MS Image Tool, to find the peaks in our LA-ICP-MS data [16]. Once the peak areas have been found, the %RSD of the data can be calculated to provide information on the variability in label uptake across the cell population. We find that plotting a histogram of the integrated signals is a useful way of visually displaying this information in publications and conference presentations. For an example histogram, see reference [10].
16. Some modern systems have updates that allow the option of sending a trigger pulse before every laser shot. This approach may simplify your data processing. It is advisable to speak to your relevant LA and ICP-MS service engineers to discuss whether this is a possible option for your instrumentation.

References

1. Giesen C, Wang HA, Schapiro D, Zivanovic N, Jacobs A, Hattendorf B, Schuffler PJ, Grolimund D, Buhmann JM, Brandt S, Varga Z, Wild PJ, Gunther D, Bodenmiller B (2014) Highly multiplexed imaging of tumor tissues with subcellular resolution by mass cytometry. Nat Methods 11(4):417–422. https://doi.org/10.1038/nmeth.2869
2. Theiner S, Van Malderen SJM, Van Acker T, Legin A, Keppler BK, Vanhaecke F, Koellensperger G (2017) Fast high-resolution laser ablation-inductively coupled plasma mass spectrometry imaging of the distribution of platinum-based anticancer compounds in multicellular tumor spheroids. Anal Chem 89(23): 12641–12645. https://doi.org/10.1021/acs.analchem.7b02681
3. de Vega RG, Fernandez-Sanchez ML, Pisonero J, Eiro N, Vizoso FJ, Sanz-Medel A (2017) Quantitative bioimaging of Ca, Fe, Cu and Zn in breast cancer tissues by LA-ICP-MS. J Anal Atom Spectrom 32(3): 671–677. https://doi.org/10.1039/c6ja00390g
4. Willmes M, Glessner JJG, Carleton SA, Gerrity PC, Hobbs JA (2016) Sr-87/Sr-86 isotope ratio analysis by laser ablation MC-ICP-MS in scales, spines, and fin rays as a nonlethal alternative to otoliths for reconstructing fish life history. Can J Fish Aquat Sci 73(12): 1852–1860. https://doi.org/10.1139/cjfas-2016-0103
5. Giesen C, Waentig L, Mairinger T, Drescher D, Kneipp J, Roos PH, Panne U, Jakubowski N (2011) Iodine as an elemental marker for imaging of single cells and tissue sections by laser ablation inductively coupled plasma mass spectrometry. J Anal Atom Spectrom 26(11): 2160–2165. https://doi.org/10.1039/c1ja10227c
6. Herrmann AJ, Techritz S, Jakubowski N, Haase A, Luch A, Panne U, Mueller L (2017) A simple metal staining procedure for identification and visualization of single cells by LA-ICP-MS. Analyst 142(10):1703–1710. https://doi.org/10.1039/c6an02638a
7. Mueller L, Herrmann AJ, Techritz S, Panne U, Jakubowski N (2017) Quantitative characterization of single cells by use of immunocytochemistry combined with multiplex LA-ICP-MS. Anal Bioanal Chem 409(14): 3667–3676. https://doi.org/10.1007/s00216-017-0310-1
8. Wang M, Zheng LN, Wang B, Chen HQ, Zhao YL, Chai ZF, Reid HJ, Sharp BL, Feng WY (2014) Quantitative analysis of gold nanoparticles in single cells by laser ablation inductively coupled plasma-mass spectrometry. Anal Chem 86(20):10252–10256. https://doi.org/10.1021/ac502438n
9. Drescher D, Giesen C, Traub H, Panne U, Kneipp J, Jakubowski N (2012) Quantitative imaging of gold and silver nanoparticles in single eukaryotic cells by laser ablation ICP-MS. Anal Chem 84(22):9684–9688. https://doi.org/10.1021/ac302639c
10. Managh AJ, Edwards SL, Bushell A, Wood KJ, Geissler EK, Hutchinson JA, Hutchinson RW, Reid HJ, Sharp BL (2013) Single cell tracking of gadolinium labeled CD4(+) T cells by laser ablation inductively coupled plasma mass

spectrometry. Anal Chem 85(22): 10627–10634. https://doi.org/10.1021/ac4022715

11. Managh AJ, Hutchinson RW, Riquelme P, Broichhausen C, Wege AK, Ritter U, Ahrens N, Koehl GE, Walter L, Florian C, Schlitt HJ, Reid HJ, Geissler EK, Sharp BL, Hutchinson JA (2014) Laser ablation-inductively coupled plasma mass spectrometry: an emerging technology for detecting rare cells in tissue sections. J Immunol 193(5): 2600–2608. https://doi.org/10.4049/jimmunol.1400869
12. Reifschneider O, Wentker KS, Strobel K, Schmidt R, Masthoff M, Sperling M, Faber C, Karst U (2015) Elemental bioimaging of thulium in mouse tissues by laser ablation-ICPMS as a complementary method to heteronuclear proton magnetic resonance imaging for cell tracking experiments. Anal Chem 87(8): 4225–4230. https://doi.org/10.1021/ac504363q
13. Douglas DN, Managh AJ, Reid HJ, Sharp BL (2015) High-speed, integrated ablation cell and dual concentric injector plasma torch for laser ablation-inductively coupled plasma mass spectrometry. Anal Chem 87(22): 11285–11294. https://doi.org/10.1021/acs.analchem.5b02466
14. Van Malderen SJM, van Elteren JT, Vanhaecke F (2015) Development of a fast laser ablation-inductively coupled plasma-mass spectrometry cell for sub-mu m scanning of layered materials. J Anal Atom Spectrom 30(1):119–125. https://doi.org/10.1039/c4ja00137k
15. Wang HAO, Grolimund D, Giesen C, Borca CN, Shaw-Stewart JRH, Bodenmiller B, Gunther D (2013) Fast chemical imaging at high spatial resolution by laser ablation inductively coupled plasma mass spectrometry. Anal Chem 85(21):10107–10116. https://doi.org/10.1021/ac400996x
16. Reid P, Managh AJ (2017) LA-ICP-MS ImageTool. https://www.reid-it.co.uk/downloads/down_la_icp_ms_it.htm
17. Managh AJ, Douglas DN, Cowen KM, Reid HJ, Sharp BL (2016) Acquisition of fast transient signals in ICP-MS with enhanced time resolution. J Anal Atom Spectrom 31(8): 1688–1692. https://doi.org/10.1039/c6ja00140h

Chapter 15

Patch-seq: Multimodal Profiling of Single-Cell Morphology, Electrophysiology, and Gene Expression

Cathryn R. Cadwell and Andreas S. Tolias

Abstract

Cells exhibit diverse morphologic phenotypes, biophysical and functional properties, and gene expression patterns. Understanding how these features are interrelated at the level of single cells has been challenging due to the lack of techniques for multimodal profiling of individual cells. We recently developed Patch-seq, a technique that combines whole-cell patch clamp recording, immunohistochemistry, and single-cell RNA-sequencing (scRNA-seq) to comprehensively profile single cells. Here we present a detailed step-by-step protocol for obtaining high-quality morphological, electrophysiological, and transcriptomic data from single cells. Patch-seq enables researchers to explore the rich, multidimensional phenotypic variability among cells and to directly correlate gene expression with phenotype at the level of single cells.

Key words Patch-seq, Electrophysiology, Single-cell RNA-sequencing, Transcriptomics, Morphology, Cell type, Whole-cell recording, Patch clamp

1 Introduction

Advances in single-cell RNA sequencing [1, 2] have enabled unbiased high-throughput molecular cell type classification in complex tissues. However, linking molecular cell types with their morphological and electrophysiological phenotypes has been difficult due to the limited availability of cell subtype–specific tools, such as viral vectors and transgenic mouse lines. Many cell types cannot be clearly defined by the differential activation of a single gene or gene regulatory element, and so new technologies that correlate transcriptomic and phenotypic data at the level of single cells provide a powerful and complementary approach to better understand the molecular underpinnings of cell type diversity.

Early attempts to combine whole-cell patch clamp recording with single-cell gene expression included reverse transcription polymerase chain reaction (RT-PCR) of the patch pipette contents following whole-cell recording [3–5]. However, this approach requires a priori selection of a small number of genes, providing

Miodrag Gužvić (ed.), *Single Cell Analysis: Methods and Protocols*, Methods in Molecular Biology, vol. 2752,
https://doi.org/10.1007/978-1-0716-3621-3_15,

only a glimpse of the full transcriptome and limiting the identification of new genes that may be important in defining cell types. Single-cell microarray was also used to assess genome-wide expression after patch recording [6], but microarray has a limited dynamic range, poor sensitivity, and poor specificity compared to sequencing-based approaches, and cannot detect novel transcripts or splice variants. Initial attempts to perform scRNA-seq on patched neurons yielded poor quality sequencing data [7], but we and others have shown that high-quality scRNA-seq data, comparable to that obtained from dissociated cells, can be obtained from patch clamp-recorded neurons [8, 9]. We have since further refined our methodology to improve direct morphological recovery of cells after Patch-seq [10–12]. Here we present the most up-to-date version of our Patch-seq protocol. We assume the user has a basic understanding of the patch-clamp technique and focus our instructions on modifications that are needed to facilitate Patch-seq (Fig. 1).

2 Materials

All reagents for RNase-free solutions should be ordered at the highest purity available. All glassware, spatulas, stir bars, pipettes, and pipette tips should be certified RNase-free and/or be cleaned thoroughly with RNase Zap prior to preparing any RNase-free solution.

2.1 Electrophysiology and Single-Cell RNA Sample Collection

1. Juvenile (postnatal day 15–19) or adult wild-type C57Bl/6 mice (*see* **Note 1**).
2. Isoflurane.
3. 0.2 M phosphate buffer (PB): Dissolve 23.004 g sodium phosphate dibasic (Na_2HPO_4, 142.0 g/mol) and 4.56 g sodium phosphate dibasic (NaH_2PO_4, 120.0 g/mol) in 1 L distilled water (dH_2O). Solution can be stored at room temperature (25 °C) for up to a month.
4. 0.01 M phosphate-buffered saline (PBS): Dissolve 9 g NaCl in 50 mL 0.2 M PB and 950 mL dH_2O. Solution can be stored at room temperature for up to a month.
5. RNase Zap and DNA-OFF. Wear protective gloves and eye protection when using.
6. 70% (vol/vol) ethanol: Dissolve non-molecular biology grade ethanol in dH_2O to a final concentration of 70% (vol/vol). Store in RNase Zap-treated spray bottle for up to a week.
7. Nonstick, RNase-free 1.5 mL microfuge tubes.
8. Thin-walled, RNAse-free 0.2 mL PCR tubes.

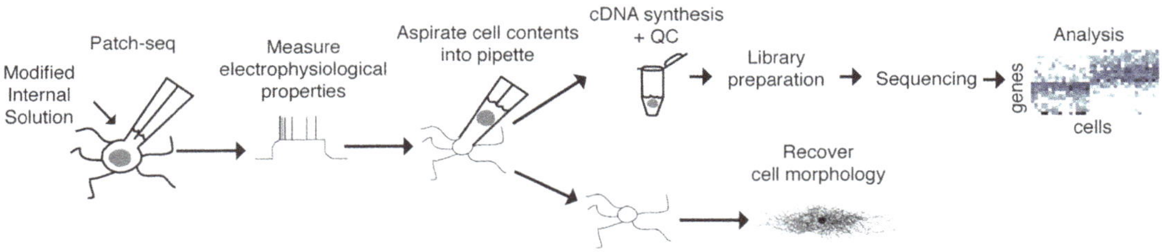

Fig. 1 Schematic of Patch-seq technique. (Adapted from [10] with permission from Springer)

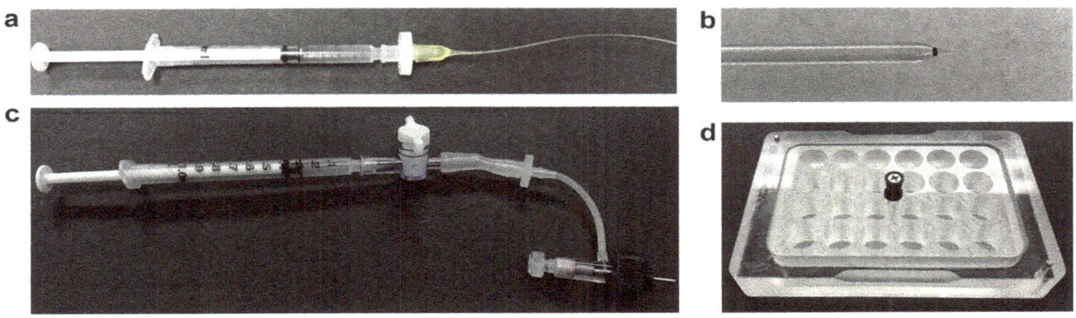

Fig. 2 Custom equipment used in Patch-seq. (**a**) 1 mL syringe with 0.2 μm syringe filter and tapered pipette tip, used for backfilling glass pipettes. (**b**) Patch pipette marked at approximately 0.3 μL volume. (**c**) Positive pressure device used to eject pipette contents into PCR tube after aspirating cell contents. (**d**) Staining chamber used for immunohistochemistry. (Reproduced from [10] with permission from Springer)

9. RNase-free oligo-$dT_{30}VN$ (sequence: 5′-AAGCAGTGGTAT CAACGCAGAG TACTTTTTTTTTTTTTTTTTTTTTTTTTTTTTTVN -3′, where V represents A, C, or G and N represents any nucleotide; HPLC purified). Dissolve in molecular biology–grade Tris-EDTA (TE) buffer at 100 μM. Store 50 μL aliquots at −20 °C for up to a year.
10. RNase-free ERCC spike-in stock solution: Thoroughly vortex ERCC RNA spike-in mix. Dilute original ERCC mix 1:4 × 10^4 in RNase-free sterile water. Store 15 μL aliquots in RNase-free 1.5 mL tubes for up to 6 months at −20 °C. Avoid freeze/thaw cycles by using a fresh aliquot for each experiment.
11. 96-well polypropylene deep well plates.
12. RNase-free triton X-100 stock: Make 33% (vol/vol) stock solution in RNase-free sterile water. Store at room temperature for up to a year. Dilute stock solution 1:100 to make working dilution 0.33% (vol/vol). Store working dilution at −20 °C for up to 6 months.
13. RNase-free 0.5 M KOH: Dissolve in RNase-free sterile water at 0.5 M. Store at room temperature for up to a year.

14. RNase-free internal solution stock: 111 mM potassium D-gluconate (K-gluconate), 4 mM potassium chloride (KCl), 10 mM HEPES, 0.2 mM ethylene glycol-bis(2-aminoethylether)-*N,N,N′,N′*-tetraacetic acid (EGTA), 4 mM adenosine 5′-triphosphate magnesium salt (Mg-ATP), 0.3 mM guanosine 5′-triphosphate sodium salt hydrate (Na-GTP), 5 mM phosphocreatine disodium salt hydrate (Na_2-phosphocreatine), 13.4 mM biocytin. First, dissolve K-gluconate KCl, HEPES, and EGTA in ~40 mL RNase-free water in a 125 mL Erlenmeyer flask. Autoclave, allow to cool to room temperature, and then add the Mg-ATP, Na-GTP, Na_2-phosphocreatine, and biocytin. Adjust the pH to 7.25 with RNase-free 0.5 M KOH using a dedicated pH meter that is cleaned with RNase Zap prior to each use. Adjust osmolarity to 235–240 mOsm using RNase-free sterile water. Filter the internal solution using an RNase-free 60 mL syringe (cleaned with RNase Zap) and 0.45 μm syringe filter. Aliquot into 1 mL samples in RNase-free 1.5 mL tubes. Store at −20 °C for up to 3 weeks.
15. Oxygenated artificial cerebrospinal fluid (ACSF): Dissolve the following salts in distilled water: 125 mM sodium chloride (NaCl), 2.5 mM potassium chloride (KCl), 1.25 mM NaH_2PO_4, 25 mM sodium bicarbonate ($NaHCO_3$), 1 mM magnesium chloride ($MgCl_2$), 11.1 mM glucose, 2 mM calcium chloride ($CaCl_2$). Oxygenate for 30 min before adding the $MgCl_2$ and $CaCl_2$. This solution can be stored at 4 °C for up to 2 weeks.
16. RNase-free lysis buffer: 0.1% (vol/vol) Triton X-100, 5 mM (each) dNTPs, 2.5 μM oligo $dT_{30}VN$, 1 U/μL recombinant RNase inhibitor, 1.6×10^{-5} ERCC (diluted from $1{:}4 \times 10^4$ ERCC stock) in nuclease-free sterile water. Thaw all reagents completely and vortex thoroughly before use. Prepare enough lysis buffer for all samples and controls immediately prior to the experiment (*see* **Note 2**).
17. 4% Paraformaldehyde (PFA): Add 40 μl of 1 N NaOH to 7.5 mL distilled water in a 50 mL Erlenmeyer flask. Cover with parafilm, and stir on a heating stir plate set on low-medium heat for 5–10 min. When the solution is warm, add 0.8 g paraformaldehyde (PFA) and re-cover with parafilm. As soon as PFA dissolves remove the solution from heat and let it cool to room temperature. Add 2 mL of 25% (wt/vol) glutaraldehyde and 10 mL of 0.2 M phosphate buffer. Swirl gently to combine. Store the PFA/glutaraldehyde solution at 4 °C. For best morphological recovery, use it the same day.
18. Recombinant RNase inhibitor (Takara Bio).

19. Slice electrophysiology rig equipped with an upright light microscope with differential interference contrast and a CCD camera.
20. Vibrating blade microtome.
21. Water bath.
22. Flaming/Brown micropipette puller with 3.0 mm square box filament. Adjust the pull setting to achieve 3–7 MΩ resistance pipettes.
23. Borosilicate glass (O.D. 2.0 mm, I.D. 1.6 mm, 10 cm length, without filament). Autoclave prior to use.
24. 1 mL syringe with tapered pipette tip, for backfilling pipettes (Fig. 2a). Clean with RNase Zap prior to use.
25. Positive pressure device for ejecting samples from pipette into PCR tubes (Fig. 2c).
26. RNase-free 50 mL conical tubes.
27. RNase-free filter pipette tips.
28. 13 mm syringe filter with 0.2 μm PTFE membrane.

2.2 cDNA Library Preparation and Sequencing

1. RNase-free template switching oligonucleotide (LNA-TSO; sequence: 5′- AAGCAGTGGTATCAACGCAGAGTACrGrG +G-3′, where rG indicates riboguanosines and +G indicates a locked nucleic acid (LNA)-modified guanosine; RNase-free; HPLC purified). Dissolve in molecular biology–grade Tris-EDTA (TE) buffer at 100 μM. Store 20 μL aliquots at −80 °C for up to a year.
2. RNase-free 5 M betaine: Dissolve in RNase-free sterile water at 5 M. Store at −20 °C for up to 6 months.
3. RNase-free reverse transcription (RT) mix: 1.75× Superscript II first strand buffer, 17.5 U/μL Superscript II reverse transcriptase, 1.75 M betaine, 8.77 mM dithiothreitol (DTT), 1.75 U/μL recombinant RNase inhibitor, 1.75 μM LNA-TSO, and 10.5 mM $MgCl_2$ in nuclease-free sterile water. Be sure that each component is completely thawed and thoroughly vortexed before adding to mix. Prepare immediately prior to the experiment or while the samples are in the thermal cycler for denaturation.
4. ISPCR Oligo (sequence: 5′- AAGCAGTGGTATCAACGCA GAGT-3′, HPLC purified). Dissolve in molecular biology–grade Tris-EDTA (TE) buffer at 100 μM. Store 50 μL aliquots at −20 °C for up to a year.
5. PCR mix: 1.67× KAPA HiFi HotStart Ready Mix, 0.17 μM ISPCR Oligo. Prepare at the time of sample processing.
6. Nonstick, RNase-free 1.5 mL microfuge tubes.

7. Nextera XT primers. Dilute the primers from the Nextera XT index kit 1:4 with sterile water. They can be stored at –20C for up to six months.
8. Tn5 transposase (produced according to protocol outlined in [13]).
9. 0.2% (wt/vol) sodium dodecyl sulfate solution (SDS): Dilute 20% (wt/vol) SDS 1:100 (vol/vol) with sterile water to make a working dilution of 0.2% (wt/vol). The solution can be stored at room temperature for up to six months.
10. 200 mM TAPS-NaOH stock solution: Dissolve TAPS in sterile molecular biology grade water and adjust pH to 8.5 (at room temperature). The solution can be stored at room temperature for up to six months.
11. 10× Tagmentation buffer: 10 mM TAPS-NaOH, 50 mM MgCl2. The solution can be stored at room temperature for up to six months.
12. Tagmentation mix: 2.5× tagmentation buffer, 25% (wt/vol) polyethylene glycol solution (PEG-8000), and 3.125 μM Tn5 transposase. Prepare immediately before use.
13. Enrichment PCR mix: 3.3× KAPA HiFi Buffer, 1 mM each dNTPs, 0.067 U/μL KAPA enzyme. Prepare immediately prior to use.
14. Thermal cycler.
15. Axygen AxyPrep mag PCR clean-up kit.
16. Ambion magnetic stand – 96.
17. 96-well polypropylene deep well plates.
18. 80% (vol/vol) ethanol: Dilute molecular biology grade ethyl alcohol in sterile water. Prepare immediately prior to use.
19. Buffer EB.
20. Agilent Bioanalyzer and High Sensitivity DNA kit.
21. Qubit 3.0 quantitation kit.
22. A compatible sequencing machine.

2.3 Immunohistochemistry and Morphological Recovery

1. 24-well tissue culture plates.
2. Custom staining chamber (Fig. 2d).
3. Clinical lab rotator set to low (~60–80 rpm).
4. 0.01 M phosphate-buffered saline (PBS): Dissolve 9 g NaCl in 50 mL 0.2 M PB and 950 mL dH_2O. Solution can be stored at room temperature for up to a month.
5. 3% (vol/vol) hydrogen peroxide (H_2O_2): Combine 45 mL of 0.01 M PBS with 5 mL of 30% (vol/vol) H_2O_2. Vortex briefly. Prepare immediately prior to use.

6. Vectastain ABC-HRP Elite Kit (Standard).
7. A/B mix: Combine 10 mL of 0.01 M PBS, 100 μL Triton X-100, 4 drops of reagent A, and 4 drops of reagent B (from ABC Elite Kit) in a 50 mL conical tube. Vortex briefly. Prepare immediately prior to use.
8. 3,3′-diaminobenzidine (DAB) stock solution: Dissolve DAB powder at 20 mg/mL in DMSO. Store aliquots at −20 °C for up to 1 year.
9. DAB mix: Combine 10 mL 0.01 M PBS, 0.25 mL DAB stock solution, and 3.3 μL 30% (vol/vol) H_2O_2 in a 50 mL conical tube. Vortex briefly. Prepare immediately prior to use.
10. 0.2 M Tris buffer: Dissolve 2.4228 g Trizma base (121.14 g/mol) in 35–50 mL dH2O. Adjust pH to 8.5 using 1 M HCl. Add dH2O to total volume of 100 mL. Solution can be stored at room temperature for up to 6 months.
11. Mowiol mounting medium: Weigh out 6 g glycerol in a 50 mL conical tube. Add 2.4 g mowiol 4–88 and vortex thoroughly. Add 6 mL dH2O, vortex, and leave at room temperature for 2 h. Add 12 mL 0.2 M Tris buffer and incubate at ~53 °C, vortexing occasionally, until the mowiol dissolves. Clarify by centrifugation at 2000–3000*g* for 20 min. Store 1 mL aliquots in 1.5 mL tubes at −20 °C for up to 1 year. Once thawed, the solution is stable at room temperature for up to a month.
12. Superfrost plus gold microscope slides.
13. Cover slips.
14. Light microscope for visualizing cell morphology.

3 Methods

3.1 Electrophysiology and Single-Cell RNA Sample Collection

1. Prepare all work surfaces and equipment that may come in contact with RNase-free solutions by cleaning with RNase Zap, followed by 70% (vol/vol) ethanol (*see* **Note 3**). Any equipment that may have come into contact with post-PCR products (tube holders, pipettes) should also be cleaned thoroughly with DNA-OFF before RNase Zap. Items can be cleaned the day before the experiment.
2. Thaw an aliquot of RNase-free internal solution stock, and add recombinant RNase inhibitor to a final concentration of 1 U/μL and vortex well. Addition of RNase inhibitor should increase the osmolarity of the internal solution to the desired range of 315–320 mOsm. Adjust the osmolarity of the ACSF to be 18–20 mOsm lower than the internal solution, adding small amounts of sucrose as needed to increase the osmolarity of the ACSF (*see* **Note 4**).

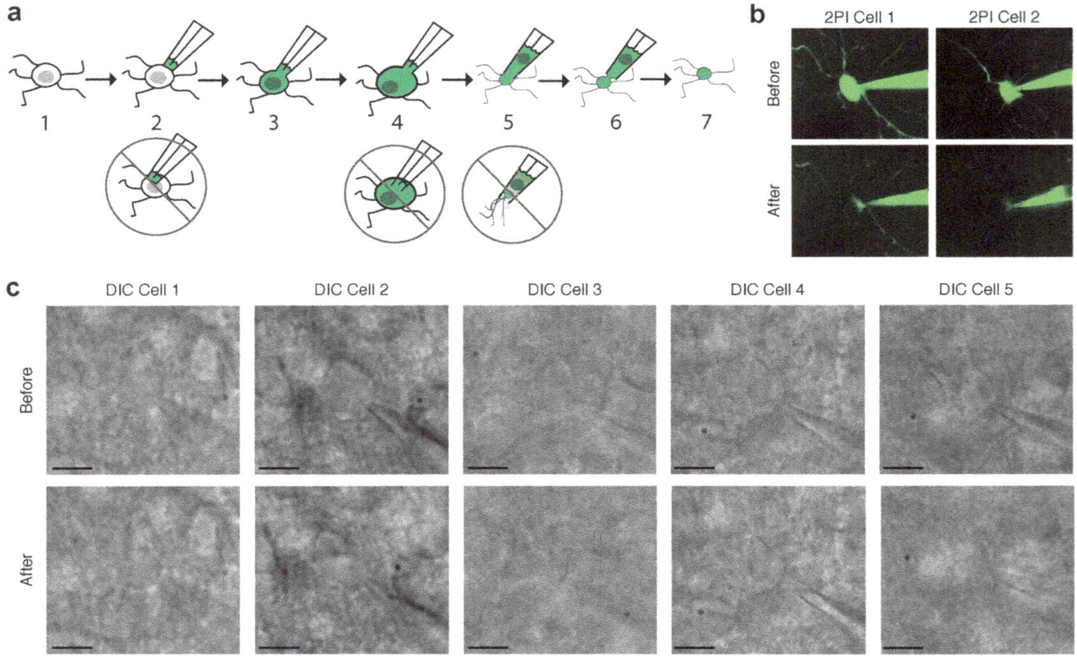

Fig. 3 Ideal patching approach for aspiration of cell contents. (**a**) The target cell is approached from the side rather than the top to provide the most direct path for cell contents to enter the pipette. As the cell slowly swells, it may be helpful to move the pipette back a few microns to release the tension on the membrane and prevent current leakage. During aspiration of the cell contents, the cell is continuously monitored to ensure that no extracellular contents are entering the pipette. Following aspiration, the pipette is slowly pulled away from the cell membrane, leaving the shrunken cell body intact in the tissue slice to preserve axodendritic morphology. (**b**) Examples of collapse of the cell body after aspiration of cell contents, visualized using two-photon imaging (2PI). Scale bars, 10 μm. (**c**) Examples of collapse of the cell body after aspiration of cell contents, visualized using differential interference contrast (DIC) imaging. Scale bar, 10 μm. (Reproduced from [10] with permission from Springer)

3. Draw the internal solution with RNase inhibitor into a 1 mL syringe, and push through a 0.2 μm syringe filter into the tapered pipette tip (Fig. 2a, *see* **Note 5**).
4. Prepare sample tubes by adding 4 μL of lysis buffer to each RNase-free 0.2 mL PCR tube. Store tubes in a clean (RNase Zap–treated) tube holder on ice.
5. Deeply anesthetize the mouse with isoflurane and cut acute brain slices according to standard protocols for the brain region of interest (e.g., [11, 14]).
6. Allow the slices to recover in oxygenated ACSF for 1 h at room temperature (~25 °C).
7. Place a tissue slice in the recording chamber and select a cell to target using differential interference contrast (DIC) and/or wide-field fluorescence light microscopy.

8. Pull an autoclaved glass pipette and backfill with 0.1–0.3 μL of internal solution (Fig. 2b). Tap or shake to remove any bubbles (*see* **Note 6**).
9. Secure the glass pipette in the pipette holder and lower it into the circulating ACSF. Immediately apply a small amount of positive pressure and push out any air bubbles. Measure the resistance (*see* **Note 7**).
10. Advance the pipette up to the target cell, maintaining a small amount of positive pressure, until the cell membrane is observed to dimple under the tip and the resistance across the pipette slightly increases.
11. Release the positive pressure and apply a small amount of suction, if necessary, to achieve a GΩ seal on the cell membrane. Set the holding potential to −70 mV and compensate the fast capacitance. Reposition the pipette if needed so that the cell membrane is not under tension, and the pipette approaches from the side of the cell rather than the top (Fig. 3).
12. Apply brisk suction to achieve whole-cell access (*see* **Note 8**). Compensate the slow capacitance.
13. Run the desired electrophysiology protocol, typically a series of hyperpolarizing, subthreshold, and suprathreshold current steps to record the intrinsic biophysical properties and firing pattern of the cell.
14. If morphological recovery is desired, hold the cell in current clamp mode for 30–60 min (*see* **Note 9**).
15. Aspirate the cell contents into the patch pipette by gently applying negative pressure (0.7–1.5 psi, *see* **Note 10**). Monitor the holding currents and size of the cell body on DIC. If the cell becomes unstable, discontinue the suction and wait a few minutes to see if it recovers. Continue until the cell body has significantly decreased in size (typically 2–10 min) or the cell cannot tolerate more suction.
16. Release the suction and remove the pipette (*see* **Note 11**).
17. Immediately eject the contents of the pipette into an RNase-free 0.2 mL PCR tube containing 4 μL lysis buffer (prepared in step 4) by applying a small amount of positive pressure (using the device depicted in Fig. 2c, or something similar) and gently breaking the tip on the bottom of the tube. Avoid making any bubbles. Cap the PCR tube and store on ice until all samples have been collected.
18. If morphological recovery is desired, immediately remove the slice from the recording chamber and place it in one well of a 24-well culture dish. Remove the excess ACSF and add ~1 mL of cold, fresh PFA. Cover the plate with parafilm underneath the lid to form an airtight seal (*see* **Note 12**).

3.2 cDNA Library Preparation and Sequencing

1. Remove any remaining bubbles by gently tapping the side of each tube. Spin down the samples at 700 g for 10 s at 4 °C and place back on ice.
2. Denature the RNA by incubating the samples in a thermal cycler for 3 min at 72 °C.
3. Add 5.7 μL of RT mix to each tube and pipette up and down four times to mix. Gently tap each tube as needed to remove bubbles and spin down at 700 g for 10 s at 4 °C. Immediately return to ice.
4. Incubate the samples in a thermal cycler according to the following program: **step 1**: 42 °C for 90 min; **step 2**: 50 °C for 2 min; **step 3**: 42 °C for 2 min (repeat **steps 2–3** ten times); **step 4**: 70 °C for 10 min; **step 5**: Hold at 4 °C (*see* **Note 13**).
5. Add 15 μL of PCR mix to each tube, vortex to mix, and spin down at 700 g for 10 s at 4 °C.
6. Incubate the samples in a thermal cycler according to the following program: **step 1**: 98 °C for 3 min; **step 2**: 98 °C for 20 s; **step 3**: 67 °C for 15 s; **step 4**: 72 °C for 6 min (repeat **steps 2–4** eighteen times); **step 5**: 72 °C for 5 min; **step 6**: Hold at 4 °C (*see* **Notes 14** and **15**).
7. Allow the Axygen AxyPrep mag PCR beads to come to room temperature for 15 min. Vortex well.
8. Add 17.5 μL of beads to each sample and mix thoroughly by pipetting up and down at least 10 times.
9. Transfer samples to a 96-well plate, cover with lid, and incubate for 8 min at room temperature.
10. Place the 96-well plate on the magnetic stand and incubate for an additional 5 min.
11. Carefully remove the clear solution without disturbing the beads.
12. Wash the beads two times with 80% ethanol for 30 s each time. Carefully remove the ethanol without touching the beads after each wash.
13. Cover the plate with a lid and leave at room temperature until the beads are completely dry and small cracks appear on the surface (typically 5–10 min).
14. Take the plate off the magnetic stand, add 17.5 μL of Buffer EB to each well, and pipette up and down ~10 times to resuspend the beads.
15. Incubate the plate off the magnetic stand for 2 min.
16. Place the plate back on the magnetic stand and incubate for an additional 2–3 min. Without touching the beads, collect 15 μL

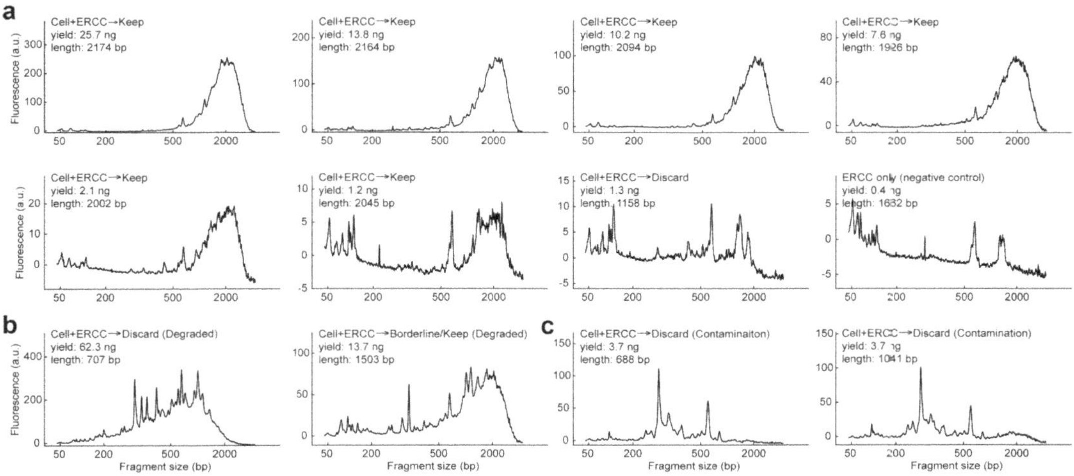

Fig. 4 Quality check of full-length single-cell cDNA libraries. Bioanalyzer profiles of single-cell amplified cDNA from several cells from a typical experiment (**a**) showing variable cDNA yield, as well as a negative control from the same experiment that consisted of ERCC spike-in RNA only. Profiles from degraded (**b**) or contaminated (**c**) samples are also shown. The low molecular weight fragments in (**c**) were traced back to bacterial contamination of a reagent. (Reproduced from [10] with permission from Springer)

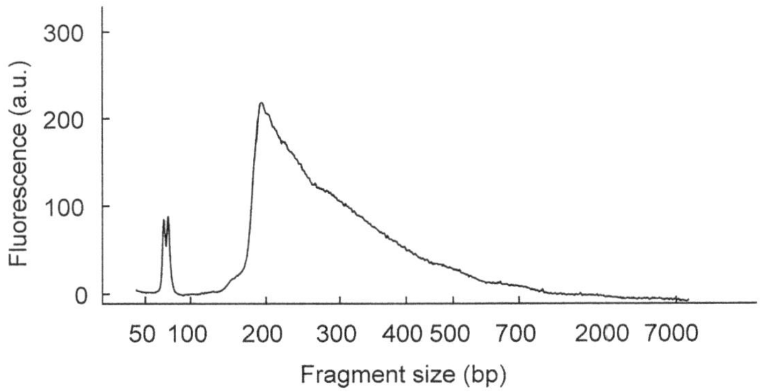

Fig. 5 Quality check of pooled sequencing library. Example of bioanalyzer profile showing the size distribution of the final pooled sequencing library. (Reproduced from [10] with permission from Springer)

of the clear liquid from each well into a new 0.2 mL PCR tube (*see* **Note 16**).

17. Check the concentration and size distribution of the full-length cDNA library on an Agilent Bioanalyzer using the Agilent High Sensitivity DNA Kit according to the manufacturer's instructions (Fig. 4; *see* **Note 17**).
18. Dilute each sample to 50 pg/μL.
19. Add 4 μL of tagmentation mix to 6 μL of each sample. Vortex well.

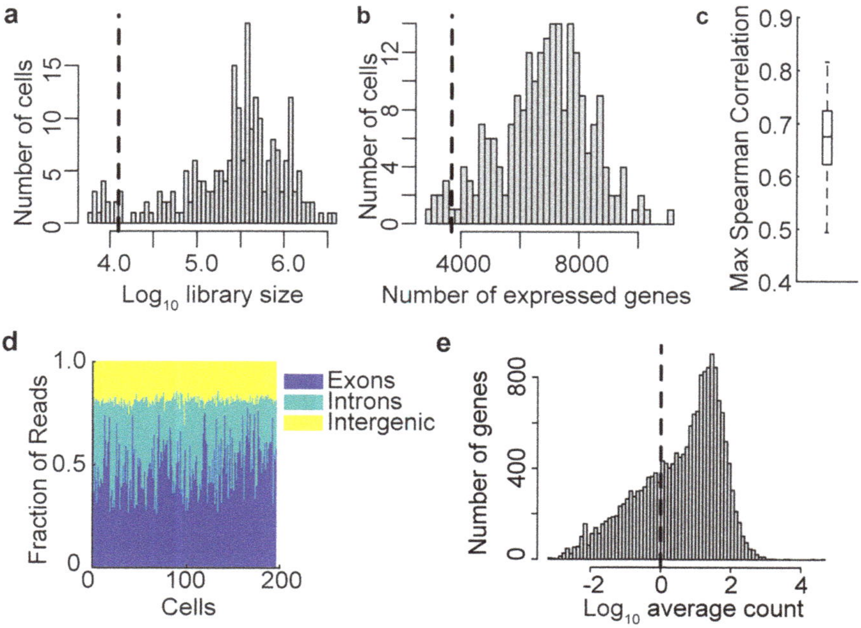

Fig. 6 Post-sequencing quality control. The quality of the samples can be assessed by analyzing the sequencing depth (**a**), number of detected genes (**b**), the maximum spearman correlation between each cell and all other cells (**c**), and fraction of reads mapped to exons, introns, and intergenic segments (**d**). For this data set approximately 10% of the samples were discarded post-sequencing based on pre-determined quality control criteria (dashed lines in (**a**) and (**b**) represent the cutoffs below which samples were discarded). (**e**) Histogram of mean normalized expression across cells for all genes in the example dataset. Genes with a mean expression of less than one (dashed line) can be excluded to minimize the impact of gene dropout on downstream analyses. (Reproduced from [10] with permission from Springer)

20. Incubate the samples in a thermal cycler at 55 °C for 8 min.
21. Add 2.5 μL of 0.2% SDS to each sample and incubate at room temperature for 5 min. Return to ice.
22. Add 2.5 μL each of index 1 (N7XX) and index 2 (S5XX) primers, ensuring that each sample receives a unique combination of indices.
23. Add 7.5 μL of enrichment PCR mix to each tube, vortex and spin down at 700 g for 10 s at 4 °C.
24. Incubate the samples in a thermal cycler according to the following program: **step 1**: 72 °C for 3 min; **step 2**: 95 °C for 30 s; **step 3**: 95 °C for 10 s; **step 4**: 55 °C for 30 s; **step 5**: 72 °C for 30 s (repeat **steps 3–5** 12 times); **step 6**: 72 °C for 5 min; **step 7**: Hold at 4 °C.
25. Pool 2.5 μL in a 1.5 mL tube.
26. Perform PCR purification on the pooled, amplified cDNA as described above (steps 7–16) with the following changes: 1) use a 1:1 ratio of beads to cDNA, 2) resuspend the beads in

0.875 µL × the number of samples, 3) collect a final volume of 0.75 µL × the number of samples.

27. Determine the mean size and concentration of the final cDNA library using an Agilent Bioanalyzer and Qubit, respectively (Fig. 5; *see* **Note 18**). Dilute the library to ~3 nM.
28. Sequence the library on a compatible machine, such as Illumina NextSeq 500, according to the manufacturer's instructions.
29. Demultiplex the reads and align to the appropriate reference genome.
30. Assess the quality of the samples and exclude poor quality samples using predetermined criteria (Fig. 6).
31. Perform additional exploratory analyses as desired (*see* **Note 19**).

3.3 Immunohistochemistry and Morphological Recovery

1. After 10–14 days of fixation, transfer the tissue slices to a staining chamber with mesh-bottom wells (Fig. 1d, *see* **Note 20**).
2. Carefully remove any trace of PFA and wash the slices three times with 0.01 M PBS. Cover and leave on the rotator for 10 min at room temperature during each wash.
3. Remove the PBS and submerge slices in the 3% (vol/vol) H_2O_2. Cover and leave on the rotator at room temperature for 30 min.
4. Remove the 3% (vol/vol) H_2O_2 and wash the slices three times with 0.01 M PBS. Cover and leave on the rotator for 10 min at room temperature during each wash.
5. Remove the PBS and add the A/B mix. Cover and store overnight at 4 °C.
6. Remove the A/B mix and wash the slices four times with 0.01 M PBS. Cover and leave on the rotator for 10 min at room temperature during each wash.
7. Remove the PBS and add the DAB mix. Leave on the rotator at room temperature for 7–8 min.
8. Remove the DAB mix and wash the slices three times with 0.01 M PBS. Cover and leave on the rotator for 10 min at room temperature during each wash.
9. Mount the slices on microscope slides in Mowiol mounting medium and cover with a glass coverslip (*see* **Note 21**).
10. Visualize cell morphology under a light microscope (Fig. 7, *see* **Note 22**).

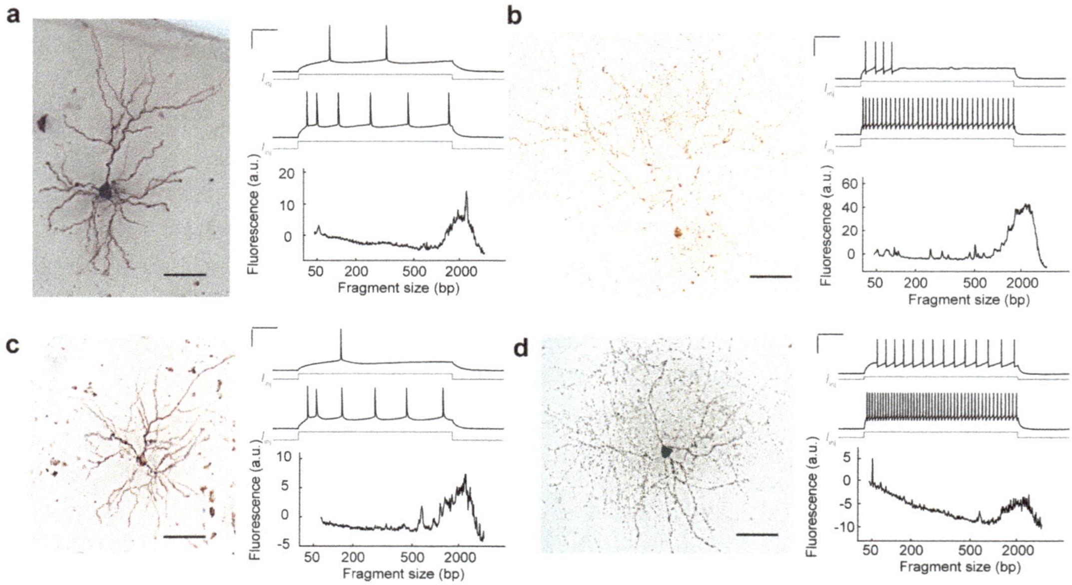

Fig. 7 Direct morphological recovery of Patch-seq cells. Examples of two excitatory (**a**, **c**) and two inhibitory (**b**, **d**) neurons with successful characterization of morphologic and electrophysiological profiles as well as acquisition of full-length cDNA. Scale bars for histochemical staining, 50 μm. Scale bars for electrophysiological traces, 100 ms and 50 mV/500 pA. (Adapted from [10] with permission from Springer)

4 Notes

1. Experiments involving animals should be carried out in accordance with the National Research Council's "Guide for the Care and Use of Laboratory Animals" and with approval from the appropriate Institutional Animal Care and Use Committee.
2. Inclusion of appropriate positive and negative controls early in the sample collection process, and at various intermediate stages of processing, is critical to ensure sample quality and can help tremendously to localize problems if sample quality becomes compromised. Endogenous (within the tissue itself) or exogenous (from the environment or experimenter) RNase, RNA from cells other than the one targeted, and cross-contamination with amplified cDNA from prior experiments are all important potential sources of contamination that should be eliminated to the extent possible and controlled for by experimental design. Another important consideration when designing a Patch-seq experiment is technical variability and bias that can be introduced during the library preparation, PCR amplification, and sequencing. To control for this, it is helpful to collect and process samples in a randomized fashion if possible and to assess results for batch effects.

3. Great care should be taken to reduce or eliminate any source of RNase contamination during sample collection and solution preparation. Items should be re-cleaned during the experiment if they may have become contaminated (e.g., if the glass pipette breaks and the wire electrode comes into contact with ACSF). In addition to using sterile techniques, certified RNase-free solutions, tubes, and pipette tips should be used whenever possible. Any reagents not available as certified RNase-free should be ordered at the highest purity available. Glassware, pipettes, counters, and all other equipment should be cleaned thoroughly with RNase Zap and autoclaved if possible. Ideally, a dedicated RNase-free positive pressure room or dedicated RNase-free bench stocked with RNase-free equipment (pipettes, tips, glassware, tubes, vortex, tabletop centrifuge) should be used exclusively for pre-RT sample handling steps.
4. Let the ACSF and internal solution come to room temperature and alternate measuring the osmolarity of the two solutions to obtain the most accurate comparative measurements. Add small amounts of water or sucrose and stir thoroughly before each measurement. Obtain at least three measurements when approaching the desired osmolarity (within 10 mOsm of the target osmolarity) to ensure the readings are stable before making further adjustments.
5. Carefully tap the syringe to remove any air bubbles before pushing the internal solution through the filter. If not removed, air bubbles will destroy the integrity of the syringe filter and can lead to clogging of the pipette during patching.
6. Using too much internal solution will inhibit the downstream reverse transcription reaction.
7. In general, pipettes of 3–4 MΩ are typically ideal for larger cells such as layer V pyramidal neurons, and pipettes of 5–7 MΩ are ideal for smaller cells such as interneurons.
8. Discontinue suction immediately when the cell membrane opens to prevent losing the seal between the cell membrane and the pipette. Signs of a broken seal include an unstable resting membrane potential, a large current required to hold the cell at −70 mV (>200–300 pA), or blunted/absent action potentials when stimulated in the current clamp. If the recording seems unstable or there is a large amount of current leakage, try adjusting the pipette position or wait a few minutes and, if there is no improvement, then the sample should be discarded, and a new cell should be targeted.
9. Injecting the cell with large depolarizing currents (e.g., 15 × 100 ms pulses of 10–20 nA delivered at 1 Hz) during this holding period may help to diffuse biocytin throughout the fine axonal processes and improve morphological recovery.

10. Be cautious to apply the suction gently to avoid losing the seal. Use a large (20–30 mL), new syringe that is partially withdrawn, and ensure smooth contact between the rubber piston and syringe wall. If the cell is lost (i.e., there is a sudden increase in the amount of current required to hold the cell at −70 mV), immediately discontinue suction and remove the pipette. The RNA may still be salvageable, but endogenous RNases from the extracellular space will degrade it quickly. If a substantial amount of extracellular material has entered the pipette, the sample should be discarded.
11. If attempting to preserve cell morphology, first move the pipette back slowly from the cell and allow the membrane to slowly retract from the pipette before removing the pipette from the tissue.
12. The fixed tissue slice can be stored at 4 °C for 10–14 days before proceeding with immunohistochemistry.
13. Samples can be left in the PCR machine overnight while running the RT reaction.
14. PCR Products can be stored at −20 or −80 °C for up to 6 months before proceeding.
15. Amplified cDNA should be handled and stored in a separate post-PCR area away from the pre-PCR samples to prevent cross-contamination of new samples with amplified cDNA.
16. Purified full-length cDNA can be stored at −20 or −80 °C for up to 6 months before proceeding with tagmentation and sequencing.
17. The full-length cDNA should have a broad peak at 1500–2000 bp and very few fragments <500 bp.
18. The mean length of fragments in the final sequencing library should be <500 bp.
19. A complete overview of possible analytical approaches is beyond the scope of this chapter. In addition to the repertoire of scRNA-seq analytical tools, additional resources specific to analyzing Patch-seq data sets include [15–17].
20. If a specialized staining chamber is not available, staining can be performed in a 24-well plate by changing the solution in each well separately.
21. The microscope slides can be stored indefinitely at room temperature, protected from light.
22. Staining quality may continue to improve for 48–72 h after mounting on slides.

References

1. Picelli S, Bjorklund AK, Faridani OR, Sagasser S, Winberg G, Sandberg R (2013) Smart-seq2 for sensitive full-length transcriptome profiling in single cells. Nat Methods 10(11):1096–1098. https://doi.org/10.1038/nmeth.2639
2. Tang F, Barbacioru C, Wang Y, Nordman E, Lee C, Xu N, Wang X, Bodeau J, Tuch BB, Siddiqui A, Lao K, Surani MA (2009) mRNA-Seq whole-transcriptome analysis of a single cell. Nat Methods 6(5):377–382. https://doi.org/10.1038/nmeth.1315
3. Eberwine J, Yeh H, Miyashiro K, Cao Y, Nair S, Finnell R, Zettel M, Coleman P (1992) Analysis of gene expression in single live neurons. Proc Natl Acad Sci U S A 89(7):3010–3014. https://doi.org/10.1073/pnas.89.7.3010
4. Sucher NJ, Deitcher DL (1995) PCR and patch-clamp analysis of single neurons. Neuron 14(6):1095–1100. https://doi.org/10.1016/0896-6273(95)90257-0
5. Sucher NJ, Deitcher DL, Baro DJ, Warrick RM, Guenther E (2000) Genes and channels: patch/voltage-clamp analysis and single-cell RT-PCR. Cell Tissue Res 302(3):295–307. https://doi.org/10.1007/s004410000289
6. Subkhankulova T, Yano K, Robinson HP, Livesey FJ (2010) Grouping and classifying electrophysiologically-defined classes of neocortical neurons by single cell, whole-genome expression profiling. Front Mol Neurosci 3:10. https://doi.org/10.3389/fnmol.2010.00010
7. Qiu S, Luo S, Evgrafov O, Li R, Schroth GP, Levitt P, Knowles JA, Wang K (2012) Single-neuron RNA-Seq: technical feasibility and reproducibility. Front Genet 3:124. https://doi.org/10.3389/fgene.2012.00124
8. Cadwell CR, Palasantza A, Jiang X, Berens P, Deng Q, Yilmaz M, Reimer J, Shen S, Bethge M, Tolias KF, Sandberg R, Tolias AS (2016) Electrophysiological, transcriptomic and morphologic profiling of single neurons using patch-seq. Nat Biotechnol 34(2):199–203. https://doi.org/10.1038/nbt.3445
9. Fuzik J, Zeisel A, Mate Z, Calvigioni D, Yanagawa Y, Szabo G, Linnarsson S, Harkany T (2016) Integration of electrophysiological recordings with single-cell RNA-seq data identifies neuronal subtypes. Nat Biotechnol 34(2):175–183. https://doi.org/10.1038/nbt.3443
10. Cadwell CR, Scala F, Li S, Livrizzi G, Shen S, Sandberg R, Jiang X, Tolias AS (2017) Multimodal profiling of single-cell morphology, electrophysiology, and gene expression using Patch-seq. Nat Protoc 12(12):2531–2553. https://doi.org/10.1038/nprot.2017.120
11. Scala F, Kobak D, Bernabucci M, Bernaerts Y, Cadwell CR, Castro JR, Hartmanis L, Jiang X, Laturnus S, Miranda E, Mulherkar S, Tan ZH, Yao Z, Zeng H, Sandberg R, Berens P, Tolias AS (2021) Phenotypic variation of transcriptomic cell types in mouse motor cortex. Nature 598(7879):144–150. https://doi.org/10.1038/s41586-020-2907-3
12. Scala F, Kobak D, Shan S, Bernaerts Y, Laturnus S, Cadwell CR, Hartmanis L, Froudarakis E, Castro JR, Tan ZH, Papadopoulos S, Patel SS, Sandberg R, Berens P, Jiang X, Tolias AS (2019) Layer 4 of mouse neocortex differs in cell types and circuit organization between sensory areas. Nat Commun 10(1):4174. https://doi.org/10.1038/s41467-019-12058-z
13. Picelli S, Bjorklund AK, Reinius B, Sagasser S, Winberg G, Sandberg R (2014) Tn5 transposase and tagmentation procedures for massively scaled sequencing projects. Genome Res 24(12):2033–2040. https://doi.org/10.1101/gr.177881.114
14. Jiang X, Shen S, Cadwell CR, Berens P, Sinz F, Ecker AS, Patel S, Tolias AS (2015) Principles of connectivity among morphologically defined cell types in adult neocortex. Science 350(6264):aac9462. https://doi.org/10.1126/science.aac9462
15. Kobak D, Berens P (2019) The art of using t-SNE for single-cell transcriptomics. Nat Commun 10(1):5416. https://doi.org/10.1038/s41467-019-13056-x
16. Kobak D, Bernaerts Y, Weis MA, Scala F, Tolias AS, Berens P (2021) Sparse reduced-rank regression for exploratory visualisation of paired multivariate data. J R Stat Soc Ser C Appl Stat 70(4):980–1000. https://doi.org/10.1111/rssc.12494
17. Tripathy SJ, Toker L, Bomkamp C, Mancarci BO, Belmadani M, Pavlidis P (2018) Assessing transcriptome quality in patch-seq datasets. Front Mol Neurosci 11:363. https://doi.org/10.3389/fnmol.2018.00363

Chapter 16

Single-Nucleus ATAC-seq for Mapping Chromatin Accessibility in Individual Cells of Murine Hearts

Michail Yekelchyk, Xiang Li, Stefan Guenther, and Thomas Braun

Abstract

During the last decade a wide range of single-cell and single-nucleus next-generation sequencing techniques have been developed, which revolutionized detection of rare cell populations, enabling creation of comprehensive cell atlases of complex organs and tissues. State-of-the-art methods do not only allow classical transcriptomics of individual cells but also comprise a number of epigenetic approaches, including assessment of chromatin accessibility by single-nucleus Assay for Transposase Accessible Chromatin ATAC-seq (snATAC-seq). The snATAC-seq assay detects "open chromatin," a term for low nucleosome occupancy of genomic regions, which is a prerequisite for effective transcription factor binding. Information about open chromatin at the single-nucleus level helps to recognize epigenetic changes, sometimes before transcription of respective genes occurs. snATAC-seq detects cellular heterogeneity in otherwise still transcriptionally and/or morphologically homogeneous cell populations. Chromatin accessibility assays may be used to detect epigenetic changes in cardiac lineages during heart development, chromatin landscape changes during aging, and epigenetic alterations in heart diseases. Here, we provide an optimized protocol for snATAC-seq of murine hearts. We describe isolation of single nuclei from snap-frozen hearts, provide hints for preparation of libraries suitable for snATAC-seq next-generation sequencing (NGS) using the Chromium 10× platform, and give general recommendations for downstream analysis using conventional bioinformatic pipelines and packages. The protocol should serve as a beginner's guide to generate high-quality snATAC-seq datasets and to perform chromatin accessibility analysis of individual heart-derived cell nuclei.

Key words Heart, Single-cell, Single-nucleus ATAC-seq, Transposase, Sequencing, Chromatin accessibility, NGS, Chromium 10x™

1 Introduction

Chromatin accessibility depends on positions and density of nucleosomes, which are composed of DNA wrapped around dimers of the four core histone proteins, H2A, H2B, H3, and H4. The chromatin is often referred to as "closed" and "open," while in reality there is a gradient of chromatin accessibility from inaccessible heterochromatin to accessible euchromatin, based on the degree of nucleosome condensation [1]. The position of

Miodrag Gužvić (ed.), *Single Cell Analysis: Methods and Protocols*, Methods in Molecular Biology, vol. 2752, https://doi.org/10.1007/978-1-0716-3621-3_16,

nucleosomes is not stationary. Nucleosomes are able to slide along genomic DNA, thereby rapidly altering chromatin accessibility, either opening or closing certain regulatory regions. Nucleosome sliding or eviction allows rapid modulation of transcription factor (TF) binding, modifying gene expression according to cellular needs [2]. Different cell types acquire distinct chromatin accessibility profiles during development, which support specific transcriptional signatures [3]. Not all accessible regulatory elements are constitutively active at basal conditions. Some genes are "primed" for rapid expression when specific external or internal stimuli emerge.

Heart development goes along with alterations in chromatin accessibility as demonstrated by Jia, Preussner, and colleagues, who analyzed promoter regions of *Nkx2-5* and *Isl1*, critical genes for cardiac progenitor cells [4]. Accessibility of the *Isl1* gene primes cardiac progenitor cells to become cardiomyocytes, smooth muscles, and endothelial cells, while accessibility of the *Nkx2-5* gene restricts cardiac progenitor cells to the cardiomyocyte lineage. Single-cell (and single-nucleus) resolution allowed the authors to decipher early cell commitment steps toward different cardiac lineages in otherwise undistinguishable cardiac progenitor cells. In a nutshell, single-nucleus resolution of chromatin accessibility allows separation of distinct cell types before transcriptional changes occur [5]. It is now possible to detect branching of cell lineages at very early stages, well before the expression of specific markers of distinct cell populations commences.

The ATAC-seq method was initially developed by Jason Buenrosto in 2013 [6]. The method is based on a hyperactive Tn5 transposase, which reaches open chromatin in cell nuclei, where it cuts double-stranded DNA and simultaneously attaches specific primers to the ends of the respective fragments ("Tagmentation") [3]. Individual nuclei keep their integrity during Tagmentation and retain DNA fragments within the nuclei. This feature allows sorting of tagmented nuclei [5] and encapsulation into droplets for droplet-based single-cell/single-nucleus NGS approaches [7, 8]. In this manuscript, we provide guidance that allows users to avoid potential pitfalls and set up the snATAC-seq protocol with relatively little effort.

2 Materials

See Table 1 for the list of suggested equipment, and Table 2 for the list of necessary chemicals, reagents, and enzymes.

Table 1
Laboratory equipment

Equipment	Manufacturer
GentleMACS dissociator, M tubes	Miltenyi Biotec, Germany
40 μm cell strainer	No preference
Centrifuge (for 15 mL conical tubes and for 1.5 mL tubes)	No preference
15 mL conical tubes, 1.5 mL tubes, 0.2 mL tubes	No preference
Fluorescent-activated cell sorting (FACS) machine	No preference
Chromium Controller & Next GEM Accessory Kit	120223 or 1000202, 10xGenomics, USA
Polymerase Chain Reaction (PCR) thermocycler	No preference
Magnetic separator for SPRI bead cleanup	No preference
Qubit measurement device and HS DNA kit	Q32851, Thermo, USA
Capillary electrophoresis instrument	Agilent, USA or PerkinElmer, USA
Next Generation Sequencer	Illumina, USA

Table 2
Chemicals and enzymes

Chemicals and enzymes	Final concentration in the protocol	Manufacturer
Dulbecco's phosphate-buffered saline (DPBS)	–	Gibco / Thermo Fisher Scientific, USA
Anti-PCM-1 antibody	1:1000	HPA023374, Sigma-Aldrich, Germany
Anti-rabbit secondary antibody conjugated to Alexa 594	1:1000	A11011, Life Technologies
Chromium Next GEM Chip H Single Cell Kit	–	1000162, 10xGenomics, USA
Chromium Next GEM Single Cell ATAC Library & Gel Bead Kit v1.1	–	1000176, 10xGenomics, USA
Single Index Kit N Set A	–	1000212, 10xGenomics, USA
Nuclease-free water	–	Elut
Elution Buffer (EB)	–	Qiagen, Netherlands
Solid Phase Reversible Immobilization (SPRI) magnetic beads	–	AC-60050, MagBio, Switzerland
High-75 NextSeq500 cartridge	–	20024906, Illumina, USA

Buffer Recipes

1. *Lysis buffer:* 5 mM $CaCl_2$, 3 mM $Mg(CH_3COO)_2$, 2 mM EDTA, 0.5 mM Ethylene glycol-bis(β-aminoethyl ether)-*N*,*N*,*N*′,*N*′-tetraacetic acid (EGTA), and 10 mM Tris-HCl (pH 8).
2. *Triton buffer I*: 0.4% Triton X-100 solution: 40 μL Triton X-100 in 10 mL PBS.
3. *Triton buffer II*: 0.2% Triton X-100 solution: 20 μL Triton X-100 in 10 mL PBS.
4. *Sucrose buffer*: 1 M sucrose, 3 mM $Mg(CH_3COO)_2$, 10 mM Tris-HCl (pH 8).
5. *Staining buffer*: DPBS, 1% BSA, 0.2% Igepal CA-630, 1 mM EDTA.
6. *DAPI staining solution*: 2 ng/mL DAPI in nuclease-free water.
7. *Tagmentation Mix*: prepared in accordance with the Next GEM scATAC-seq kit manual (contains ATAC Buffer B (PN2000193) and ATAC Enzyme (PN2000123 or PN2000138)).
8. *Barcoding Mastermix*: prepared in accordance with the Next GEM scATAC-seq kit manual (contains scATAC Gel Beads v.1.1 (PN2000210), Reducing Agent B (PN2000087), Barcoding Reagent B (PN2000194), Barcoding Enzyme (PN2000125 or PN2000139), Partitioning Oil (PN2000190), and 10× controller accessory parts as per manual).
9. *Elution solution*: prepared in accordance with the Next GEM scATAC-seq kit manual (contains Buffer EB, 0.1% Tween 20 (10 μL Tween 20 in 10 mL PBS), and Reducing Agent B (PN2000087)).

3 Methods

3.1 A Brief Protocol for Cardiac Nuclei Isolation

The following protocol describes isolation of nuclei from murine hearts and optional identification of nuclei with PCM1 (pericentriolar material 1), which specifically labels cardiomyocyte nuclei [9]. Combined with fluorescence-activated cell sorting (FACS), this approach not only allows purification of all nuclei from hearts for snATAC-seq but also separation of cardiomyocytes nuclei from non-cardiomyocytes, which has advantages for some applications.

1. Sacrifice a mouse following approved procedures and dissect the heart. Place the heart into ice-cold PBS buffer to remove residual blood (*see* **Note 1**). Snap-freeze the heart in liquid

nitrogen. The frozen heart can be stored at −80 °C up to a year.

2. Thaw the frozen mouse heart in 3 mL of *Lysis buffer*. Homogenize the heart tissue utilizing Miltenyi gentleMACS™ dissociator with M tubes and the protocol "protein_01."
3. Add 3 mL of lysis buffer supplemented with *Triton buffer I* into the M tube and keep the sample on ice for 10 min.
4. Filter the cell suspension using a 40 μm cell strainer. Wash the filter with 2 mL lysis buffer, supplemented with *Triton buffer II*. Centrifuge the suspension with 1000*g* for 5 min at 4 °C.
5. Resuspend the pellet and overlay the suspension on 2 mL of *Sucrose buffer*, then centrifuge (1000*g*, 5 min, 4 °C).
6. *(Optional)* Discard the supernatant and resuspend the pellet in 500 μL of *Staining buffer*.
7. *(Optional)* Stain the nuclei with anti-PCM-1 antibody (1: 1000) for 30 min at room temperature (RT).
8. *(Optional)* Centrifuge the suspension with 1000*g* for 5 min, at 4 °C, and wash once with *Staining buffer*.
9. *(Optional)* Subsequently, incubate nuclei with an anti-rabbit secondary antibody conjugated with Alexa 594 (1:1000) for 30 min at RT. Centrifuge (1000*g*, 5 min, 4 °C) and wash once with *Staining buffer*.
10. In case the sample is not stained with the PCM1 cardiomyocyte marker (**steps 6–9**), directly resuspend the pellet from **step 5** in 500 μL *Staining buffer*. Then centrifuge the solution once with 1000*g* for 5 min at 4 °C, discard the supernatant, and resuspend the pellet again in the *Staining buffer*.
11. Stain the nuclei with *DAPI staining solution* for 10 min at RT.
12. The samples are ready for FACS. Use gating for forward and side scatter (FCC/SCC) to exclude cell debris and aggregates. Additionally, use a gate for nuclear DAPI stain (sorted nuclei should be DAPI+) (*see* **Note 2**).
13. Nuclei are sorted using the same staining buffer as in **step 10**.

3.2 Brief Description of the Chromium Next GEM Single-Cell ATAC Protocol

The following protocol briefly describes the snATAC-seq library preparation using the Chromium 10x™ kit and controller. For a detailed protocol, refer to the manufacturer's instructions [10, 11]. Here, we provide tips for library preparation from cardiac nuclei.

1. The total number of nuclei isolated from whole murine hearts is way too high for direct loading onto the Chromium 10× snATAC-seq cartridge. The nuclei suspension should be diluted. Depending on the efficiency of nuclei isolation, consecutive dilution steps may be necessary. The cartridge allows

loading of different numbers of nuclei, which adjusts the total number of data points in the analysis (*see* **Note 3**).

2. Dilute nuclei and mix them with the *Tagmentation Mix*. Incubate the mixture for 1 h at 37 °C (lid temperature = 50 °C).
3. When the incubation is done, mix tagmented nuclei with the *Barcoding Mastermix*. Next, load the Chromium 10× snATAC-seq cartridge according to the manufacturer's instructions. The separate rows are filled with gel beads, transposed nuclei in the Mastermix, and partitioning oil. Empty wells are filled with glycerol. Cover the cartridge with a rubber gasket and place in the Chromium 10× controller.
4. When the droplet generation is done, immediately transfer the acquired Gel Bead-in-Emulsion (GEMs) into a PCR strip, and run PCR cycles, using the following protocol:
   1. 72 °C 5 min
   2. 98 °C 30 s
   3. 98 °C 30 s
   4. 59 °C 30 s
   5. 72 °C 1 min
   6. go back to **step 3**, 11 times (12 cycles total)
   7. hold at 15 °C
5. After completion of the PCR, a recovery agent (see manufacturer's protocol) is added to the tubes, which breaks the droplets and forms a biphasic mixture (aqueous and pink oil fractions). The pink (oil) lower fraction is removed and the leftover solution is cleaned with Dynabeads™ MyOne SILANE beads. Finally, the DNA fragments are eluted with an *Elution solution*.
6. Clean the DNA fragments using SPRI beads (see manufacturer's protocol).
7. Cleaned DNA fragments are used for library construction. Briefly, the i5 and i7 uniquely barcoded primers, as well as the PCR Amplification mix are added to the tubes. Amplify fragments using the following PCR protocol:
   1. 98 °C 45 s
   2. 98 °C 20 s
   3. 67 °C 30 s
   4. 72 °C 20 s
   5. go back to **step 2**, 8 times (9 cycles total)
   6. 72 °C 1 min
   7. hold at 4 °C

8. Clean the library using SPRI beads according to the manufacturer's protocol.

3.3 NGS Library Quality Assessment

To check quality (QC) and size of the final snATAC-seq library, measure concentration and size distribution. We suggest to use the Qubit ™ HS DNA kit for concentration measurements (normal range for such NGS libraries is 1–50 ng/μL). Capillary electrophoresis should be employed to determine size distribution of the library (Agilent Bioanalyzer, Agilent Fragment-Analyzer, PerkinElmer Lab-Chip, or similar). We do not recommend using agarose gel electrophoresis, since its sensitivity and resolution are not sufficient for this purpose. Often snATAC-seq libraries have a "wavy structure" of size distribution, reflecting the nucleosome pattern (Fig. 1). Although a "wavy structure" is a good sign, it is not mandatory before moving forward and may depend on the cell type and state of chromatin (*see* **Note 4**).

3.4 Sequencing of the snATAC-seq Libraries

Sequencing should be performed using the Illumina platform. For example, Illumina NextSeq500 and "High-75" cartridges can be used, which allows to obtain 400–500 M of sequencing reads, accommodating one snATAC-seq library (up to 10k data points) (*see* **Note 5**). The following setup should be used for NextSeq500 (paired-end sequencing; bp):

Read 1–34;

Read 2–34;

Index 1–8;

Index 2–16.

3.5 Description of the Conventional Bioinformatical Analysis Steps

1. After sequencing, the acquired BCL output folder is used for demultiplexing of sequencing reads. Demultiplexing is done using the "*cellranger-atac mkfastq*" pipeline, which is part of the Cell Ranger ATAC analysis pipeline developed by 10xGenomics [11].
2. After the sequencing reads are demultiplexed, read mapping, peak detection, and count are accomplished via the "*cellranger-atac count*" pipeline (*see* **Note 6**).
3. *(Optional)* To combine multiple snATAC-seq runs, the "*cellranger-atac aggr*" pipeline can be used. Note, that individual runs should be first demultiplexed and counted individually (*see* **Note 7**).
4. After the counting (and aggregating, if applicable) is done, we recommend to continue the analysis in the R programming environment. We suggest to use "*Signac*" and "*Seurat*" R packages as a basis [12–14].
5. Counts and metadata are imported into R using the "*Read10x_h5*" function. The metadata and the fragments

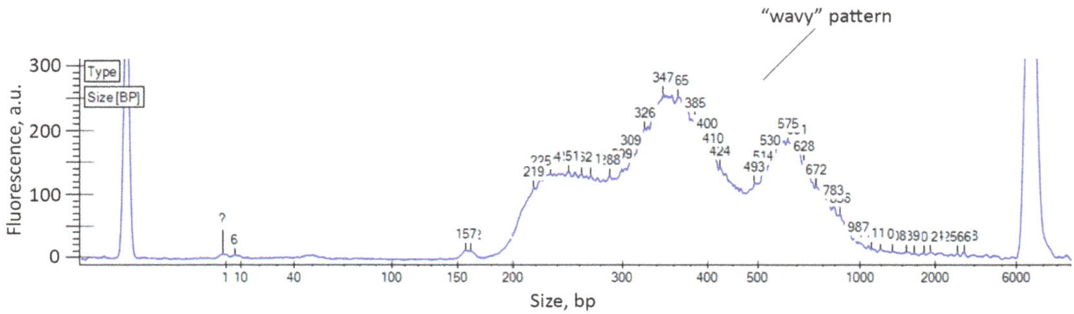

Fig. 1 Quality control of snATAC-seq libraries. A "wavy" pattern of the snATAC-seq library is a good indicator of quality. A "wavy" pattern reflects nucleosome occupancy, which may vary among different cell types. In general, a consistent "wavy" pattern is observed in cells with tight chromatin organization (e.g., various stem cell populations). Some fully differentiated somatic cells may have a less well-defined, or even absent "wavy" pattern

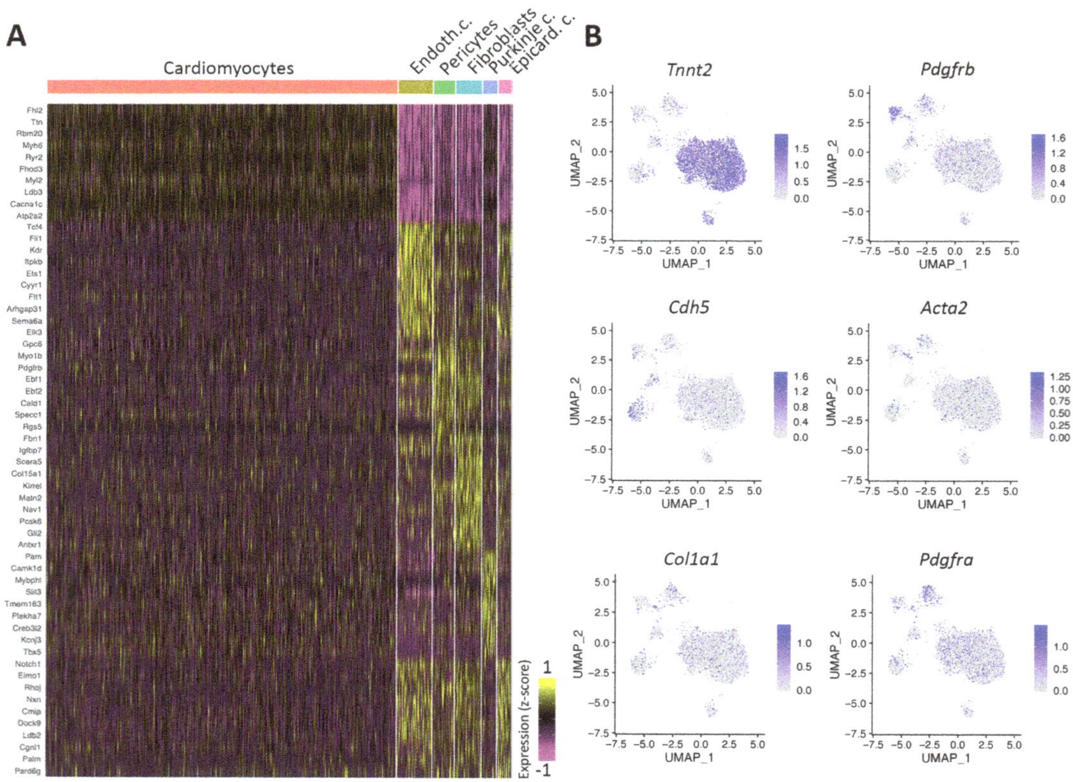

Fig. 2 Identification of different cell types in snATAC-seq datasets. (**a**) Heatmap showing chromatin accessibility ("*gene activity*") of top marker genes in different cell types of the heart. (**b**) The scatter plots overlay marker gene accessibility ("*gene activity*") in different colors over UMAP clustering (continuous scale)

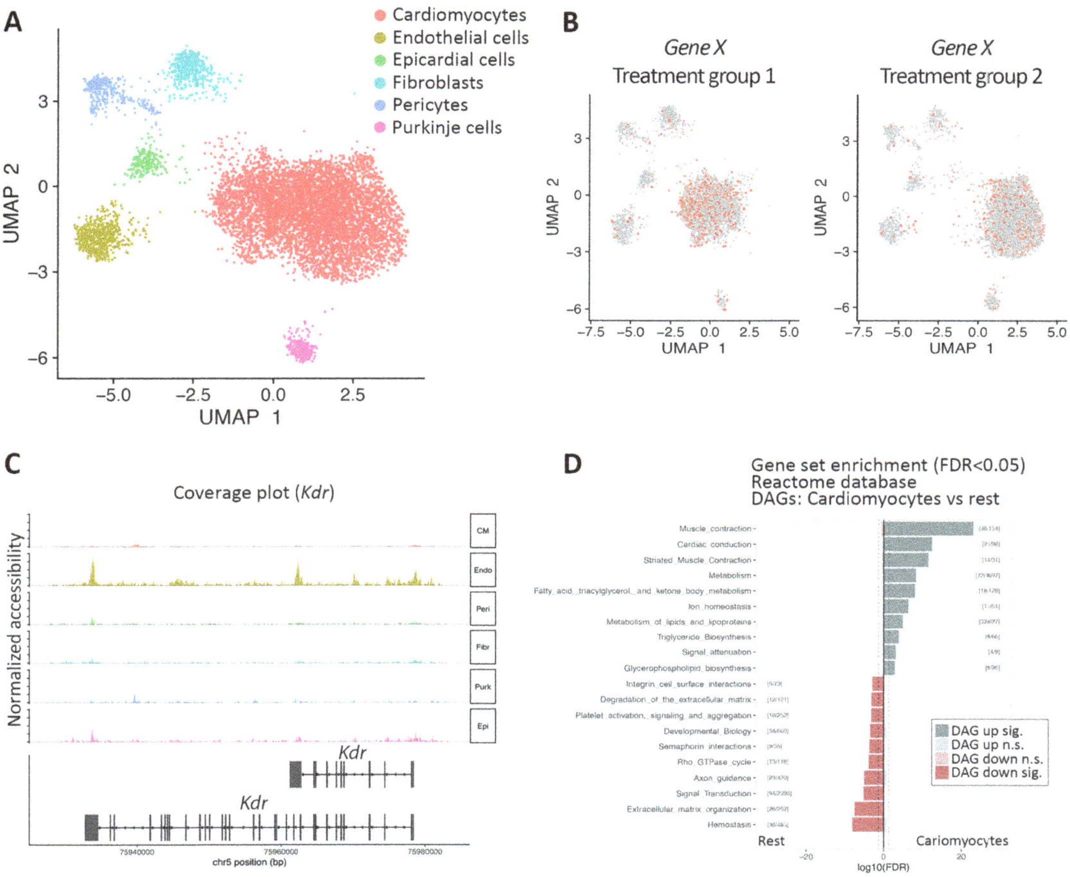

Fig. 3 Visualization of differentially accessible genes in snATAC-seq datasets. (**a**) Knowledge about marker genes accessibility profiles and their distribution across nuclei in UMAP clusters allows identification of clusters corresponding to different cell types. (**b**) Example of DAG visualization as a color overlay over UMAP clustering. (**c**) Visualization of chromatin accessibility at a gene of interest. Coverage plots are presented that show distribution of raw snATAC-seq reads across the genomic region in different UMAP clusters. (**d**) GSEA of DAGs between cardiomyocyte nuclei and all other nuclei (Reactome database)

files are read from the respective *.csv* and *.tsv.gz* files from the output folder of "*cellranger-atac count*" (or "*cellranger-atac aggr*" if applicable) pipeline.

6. The Seurat-class object is created using "*CreateSeuratObject*" function of the *Signac/Seurat* packages. The respective peak and gene annotations are acquired from the "*EnsDb.Mmusculus.v79*" (use the latest available version) package. The following QC includes the detection of nucleosome signals, transcription starting sites (TSS) enrichment, percentages of reads in the peaks, and others [15]. Based on QC evaluation, outliers are removed from the dataset.

7. Next, the dataset has to undergo normalization (based on the total number of reads per nucleus) and linear dimensional reduction (principal component analysis (PCA); or latent semantic indexing (LSI)) [16]. After completing linear dimensional reduction, nonlinear dimensional reduction has to be calculated (uniform manifold approximation and projection (UMAP)).
8. After the dimensional reduction step, the composition of the dataset and the distribution of individual nuclei can be visualised using the "*DimPlot*" function. The resulting UMAP plot will likely contain multiple "clouds" of single nuclei representing different cell types and/or cell states. An initial cluster analysis can now be performed, in particular assigning common labels to individual nuclei that share similar chromatin accessibility signatures (and hence belonging to the same UMAP "cloud"), using "*FindNeighbours*" and "*FindClusters*" functions. The clusters are therefore labeled as numbers, and many—visually— undistinguishable clusters may receive multiple internal clustering labels (*see* **Note 8**).
9. After initial clustering, a *gene activity matrix* is calculated. The matrix represents accessibility counts, associated with respective genes (we recommend to use only promoter regions: TSS + 5kb upstream). This resulting matrix represents "*gene activities*" or pseudo-expression. Of course, this only gives a hint of which genes are expressed, since gene accessibility does not accurately reflect expression.
10. After calculation of the *gene activities matrix*, we suggest to create a matrix of top "active" genes for each cluster (Fig. 2a). This allows to identify top accessibility signatures for each nuclei cluster and subsequent labeling of clusters. Accessibility of top "active" genes may be overlaid on the UMAP plot to aid definition of clusters (Fig. 2b). In our murine heart dataset, we were able to distinguish nuclei derived from cardiomyocytes, endothelial cells, epicardial cells, fibroblasts, pericytes, and Purkinje cells (Fig. 3a). The UMAP clusters are therefore renamed (and merged, if necessary) accordingly (*see* **Note 9**).
11. We recommend to apply differential accessibility analysis to obtain a list of differentially accessible genes (DAGs). This is possible using the "*MAST*" package, for example [17]. You may select the clusters of interest (e.g., cardiomyocytes vs fibroblasts, or pericytes vs the rest of the nuclei, or "young" vs "old" cardiomyocytes, or "control" vs "treated" endothelial cells, etc.) to identify DAGs among the groups (Fig. 3b).
12. DAGs are visualized as violin plots (using the "*VlnPlot*" function), or as color overlays on UMAP plots (using the "*FeaturePlot*" function). Additionally, it is possible to generate coverage

plots, which indicate the distribution of reads across the gene region of interest (Fig. 3c).

13. Finally, it is recommended to perform gene-set enrichment analysis (GSEA), using the list of previously identified DAGs, which allows to evaluate the general pattern of gene accessibility between selected clusters and/or conditions (Fig. 3d). We suggest to use different databases as reference to generate a more comprehensive picture of the results (Gene Ontology, BioCyc, Reactome, PANTHER, KEGG, etc.) [18] (*see* **Note 10**).

4 Notes

1. If the focus of the study is only on a specific part of the heart (e.g., only ventricles), parts that are not required should be removed by dissection before freezing. Avoid repetitive freezing and thawing, since each cycle disrupts chromatin organization.
2. Note that some cardiac nuclei (from cardiomyocytes) are polyploid, therefore a wide range of DAPI+ signals will occur. A wave-like pattern is expected (with peaks for 2N, 4N, 8N, etc. nuclei). A cutoff after 3–4 peaks is recommended, since otherwise it is difficult to distinguish polyploid nuclei from nuclei aggregates.
3. We suggest to aim for the maximum allowed load of the cartridge (10k data points). In our hands, the maximum capacity corresponds to 3080–7700 nuclei/μL concentration of the loading solution.
4. In case your library does not show a "wavy" pattern, you may sequence an aliquot at low sequencing depth (5–10 M reads) as a further quality control test. Since the library has a unique index, you can add a test run to a "normal" sequencing run together with another snATAC-seq sample. Evaluate the peak distribution and decide whether you want to proceed with the remaining sample.
5. If you use a high capacity Illumina sequencer (e.g., a NovaSeq™ machine), you can combine multiple snATAC-seq libraries in one run, since individual snATAC-seq libraries are uniquely indexed.
6. We recommend to use a Unix-operated server with >12 cores and >75 Gb RAM for these calculations. The longest step is the "*cellranger-atac count*," which may run for up to a few days, depending on the number of data-points and server processing power.

7. In comparison to scRNA-seq, where it is relatively straightforward to combine multiple runs in one dataset, such an operation is more difficult for snATAC-seq. The reason behind this is that a transcript annotation is fixed and therefore merging of datasets is a simple merger of matrixes, sharing the same rows (= transcripts; of note, some datasets may have transcripts that are not present in others). In case of snATAC-seq, accessibility peaks of merged fragments are being re-mapped and re-counted during the merger. Thus, new custom reference peaks are created for the combined dataset. Due to this problem, we suggest to use the "*cellranger-atac aggr*" pipeline, instead of simply merging datasets in R.
8. Automatically detected clusters often do not fully recapitulate biological differences among cell populations in the dataset. For example, multiple clusters might be identified inside one—visually—homogeneous large cluster. We suggest to analyze marker gene accessibility profiles across such potential sub-clusters and merge them into one cluster, if all cells/nuclei share the same marker gene accessibility landscape and do not own significant DAGs.
9. It is important to remember that "*gene activity*" does not mean "*expression*." Potential gene activity should not be confused with active transcription, which is detected by scRNA-seq. Some genes may be accessible, but not expressed.
10. The GSEA input is a list of genes or transcripts, which will vary according to your peak-to-gene association strategy. A broad peak-to-gene association will affect the precision of the DAGs lists and therefore GSEA. We suggest to use a relatively narrow range to associate peaks to gene promoter regions (TSS + 5 kb upstream).

Acknowledgments

This work was supported by the Excellence Initiative "Cardiopulmonary Institute" (CPI), the DFG collaborative research center SFB1213, Transregional Collaborative Research Centre 267, the DFG Transregional Collaborative Research Centre 81, and the DFG Clinical Research Unit FKO 309.

References

1. Mayran A, Drouin J (2018) Pioneer transcription factors shape the epigenetic landscape. J Biol Chem 293(36):13795–13804. https://doi.org/10.1074/jbc.R117.001232
2. Klemm SL, Shipony Z, Greenleaf WJ (2019) Chromatin accessibility and the regulatory epigenome. Nat Rev Genet 20(4):207–220. https://doi.org/10.1038/s41576-018-0089-8
3. Buenrostro JD, Wu B, Chang HY, Greenleaf WJ (2015) ATAC-seq: a method for assaying chromatin accessibility genome-wide. Curr Protoc Mol Biol 109:21.29.1–21.29.9.

https://doi.org/10.1002/0471142727.mb2129s109

4. Jia G, Preussner J, Chen X, Guenther S, Yuan X, Yekelchyk M, Kuenne C, Looso M, Zhou Y, Teichmann S, Braun T (2018) Single cell RNA-seq and ATAC-seq analysis of cardiac progenitor cell transition states and lineage settlement. Nat Commun 9(1):4877. https://doi.org/10.1038/s41467-018-07307-6
5. Chen X, Miragaia RJ, Natarajan KN, Teichmann SA (2018) A rapid and robust method for single cell chromatin accessibility profiling. Nat Commun 9(1):5345. https://doi.org/10.1038/s41467-018-07771-0
6. Buenrostro JD, Giresi PG, Zaba LC, Chang HY, Greenleaf WJ (2013) Transposition of native chromatin for fast and sensitive epigenomic profiling of open chromatin, DNA-binding proteins and nucleosome position. Nat Methods 10(12):1213–1218. https://doi.org/10.1038/nmeth.2688
7. Scott RW, Arostegui M, Schweitzer R, Rossi FMV, Underhill TM (2019) Hic1 defines quiescent mesenchymal progenitor subpopulations with distinct functions and fates in skeletal muscle regeneration. Cell Stem Cell 25(6):797–813 e799. https://doi.org/10.1016/j.stem.2019.11.004
8. Macosko EZ, Basu A, Satija R, Nemesh J, Shekhar K, Goldman M, Tirosh I, Bialas AR, Kamitaki N, Martersteck EM, Trombetta JJ, Weitz DA, Sanes JR, Shalek AK, Regev A, McCarroll SA (2015) Highly parallel genome-wide expression profiling of individual cells using nanoliter droplets. Cell 161(5):1202–1214. https://doi.org/10.1016/j.cell.2015.05.002
9. Gilsbach R, Schwaderer M, Preissl S, Gruning BA, Kranzhofer D, Schneider P, Nuhrenberg TG, Mulero-Navarro S, Weichenhan D, Braun C, Dressen M, Jacobs AR, Lahm H, Doenst T, Backofen R, Krane M, Gelb BD, Hein L (2018) Distinct epigenetic programs regulate cardiac myocyte development and disease in the human heart in vivo. Nat Commun 9(1):391. https://doi.org/10.1038/s41467-017-02762-z
10. Satpathy AT, Granja JM, Yost KE, Qi Y, Meschi F, McDermott GP, Olsen BN, Mumbach MR, Pierce SE, Corces MR, Shah P, Bell JC, Jhutty D, Nemec CM, Wang J, Wang L, Yin Y, Giresi PG, Chang ALS, Zheng GXY, Greenleaf WJ, Chang HY (2019) Massively parallel single-cell chromatin landscapes of human immune cell development and intratumoral T cell exhaustion. Nat Biotechnol 37(8):925–936. https://doi.org/10.1038/s41587-019-0206-z
11. 10xGenomics (2021) Chromium Single Cell ATAC. https://www.10xgenomics.com/products/single-cell-atac
12. Butler A, Hoffman P, Smibert P, Papalexi E, Satija R (2018) Integrating single-cell transcriptomic data across different conditions, technologies, and species. Nat Biotechnol 36(5):411–420. https://doi.org/10.1038/nbt.4096
13. Chen S, Lake BB, Zhang K (2019) High-throughput sequencing of the transcriptome and chromatin accessibility in the same cell. Nat Biotechnol 37(12):1452–1457. https://doi.org/10.1038/s41587-019-0290-0
14. Stuart T, Butler A, Hoffman P, Hafemeister C, Papalexi E, Mauck WM 3rd, Hao Y, Stoeckius M, Smibert P, Satija R (2019) Comprehensive integration of single-cell data. Cell 177(7):1888–1902. e1821. https://doi.org/10.1016/j.cell.2019.05.031
15. lab S Analyzing PBMC scATAC-seq. https://satijalab.org/signac/articles/pbmc_vignette.html
16. Tsuyuzaki K, Sato H, Sato K, Nikaido I (2020) Benchmarking principal component analysis for large-scale single-cell RNA-sequencing. Genome Biol 21(1):9. https://doi.org/10.1186/s13059-019-1900-3
17. Finak G, McDavid A, Yajima M, Deng J, Gersuk V, Shalek AK, Slichter CK, Miller HW, McElrath MJ, Prlic M, Linsley PS, Gottardo R (2015) MAST: a flexible statistical framework for assessing transcriptional changes and characterizing heterogeneity in single-cell RNA sequencing data. Genome Biol 16:278. https://doi.org/10.1186/s13059-015-0844-5
18. Subramanian A, Tamayo P, Mootha VK, Mukherjee S, Ebert BL, Gillette MA, Paulovich A, Pomeroy SL, Golub TR, Lander ES, Mesirov JP (2005) Gene set enrichment analysis: a knowledge-based approach for interpreting genome-wide expression profiles. Proc Natl Acad Sci U S A 102(43):15545–15550. https://doi.org/10.1073/pnas.0506580102

Index

Miodrag Gužvić (ed.), *Single Cell Analysis: Methods and Protocols*, Methods in Molecular Biology, vol. 2752, https://doi.org/10.1007/978-1-0716-3621-3,

P

R

S

T

V

W

Z

www.ingramcontent.com/pod-product-compliance
Ingram Content Group UK Ltd.
Pitfield, Milton Keynes, MK11 3LW, UK
UKHW051039091026
15624UKWH00011B/561
* 9 7 8 1 0 7 1 6 3 6 2 0 6 *